Mapping Innovation

Mohab Anis • Sarah Chawky • Aya Abdel Halim

Mapping Innovation

The Discipline of Building Opportunity
across Value Chains

 Springer

Mohab Anis
American University in Cairo
New Cairo, Egypt

Sarah Chawky
Innovety
Sheikh Zayed City, Egypt

Aya Abdel Halim
Innovety
Sheikh Zayed City, Egypt

ISBN 978-3-030-93629-7   ISBN 978-3-030-93627-3   (eBook)
https://doi.org/10.1007/978-3-030-93627-3

This Springer imprint is published by the registered company Springer Nature Switzerland AG
The registered company address is: Gewerbestrasse 11, 6330 Cham, Switzerland

Preface

No company is able to strive and succeed in today's market without innovation. This book acts as an eye-opener for businesses, unveiling how technology could help them redesign their products, services, and processes to differentiate themselves from competition and achieve the financial margins they desire.

The book uses simple business concepts and terminologies to help businesses realize the immense return that technology trends can bring if applied adequately. The book targets 15 industrial sectors, and in each industrial sector, there are over 10 case studies. Each case study is a standalone story that narrates in a couple of pages the potential and influence a technological innovation has on an enterprise, by defining the challenges faced, the type of technology adopted, and the impact.

We hope that companies will appreciate the power of applying technological trends and will be inspired to innovate and compete better in the market. Governments will also appreciate how technology trends are re-shaping the services they introduce to their citizens.

New Cairo, Egypt   Mohab Anis
Sheikh Zayed City, Egypt   Sarah Chawky
Sheikh Zayed City, Egypt   Aya Abdel Halim

Acknowledgements

This book emerged from our research and consultations on technological innovation over the past decade. We owe much of its inspiration to numerous people whom we interacted and learned from such as Hitendra Patel and Ron Jonash of the IXL-Center in Boston. We are also grateful to the efforts of many students at the American University in Cairo during the research phase of the book, notably Nada Afifi, Sami Mansour, and Yasmine El Wazir.

Finally, we are thankful to our editors, Charles Glaser and Arjun Narayanan, for their support and making this book possible.

Introduction

Today's world is changing rapidly, and to be able to compete effectively, the deployment of technology must be considered. Technology today has proven itself to be both beneficial and necessary and something all companies need to embrace if they wish to remain competitive in today's metamorphous marketplace. New improved technologies are revolutionizing our world and our daily lives by providing amazing resources and multi-functional devices which are now frequently considered both practical and necessary. Technology has become such an intertwined part of our lives that self-driving cars and 3D printing are no longer exciting but rather practical aspects of daily life. Robots, the Internet, and computers are now almost fixtures in daily lives, having become even faster, smaller, more efficient, and affordable than ever before.

The question that usually arises is, where exactly should I apply a technological innovation? A method that this book addresses is applying technological innovation across different parts of an industry's value chain. That way, any company would be able to assess the full extent a technology's potential across all its function phases, for example, design, development, distribution, sales, and marketing. Every time a technology is applied to a part within an industry's value chain, it would potentially bring value to the topline and/or bottom line.

Contents

About the Authors

Mohab Anis is Professor of Engineering at the American University in Cairo; CEO of INNOVETY, a leading innovation management consulting firm; and principal of 1XL Center North Africa & Levant. Mohab loves helping companies build new business opportunities and has consulted widely in the areas of innovation strategy. This includes auditing organizations' innovation capabilities, intellectual property strategies, business planning, startup and SME operations to scale up, technology transfer, and go-to-market strategies. Mohab consulted with more than 300 businesses covering more than 10 sectors in 15 countries, mostly in MENA. He spearheaded more than 20 innovation and industrial strategies for 5 MENA governments. In 2018, he overlooked the ICT for development roadmap for Swiss Terre des Hommes in Egypt, Lebanon ,and Jordan.

Earlier, Mohab spent years as a tenured Professor of Electrical and Computer Engineering at University of Waterloo, Canada, to which he's now adjunct. There, he consulted, in the USA and Canada, to a variety of Fortune 500 companies, and was also involved with Waterloo's technology transfer office where he worked on identifying technologies that have the highest potential for commercialization.

As an academic, he has authored 170 international papers, 5 books, and 3 US patents. He has been on the editorial board of 10 international journals and has supervised 15 PhD and 16 master's students. Mohab was awarded three of Canada's highest awards for excellence in innovation (The Early Research Award from Ontario's Ministry of Research and Innovation, the Colton Medal for Research Excellence, and the IEEE International Design Award), as well as both of AUC's top awards in Teaching and Research & Creative Endeavours. At the AUC, Mohab teaches electronics and VLSI design, advanced microelectronics systems, innovation strategy, economics and management of nanotechnology, business consulting, and introduction to business.

Mohab holds a PhD in computer engineering from the University of Waterloo (2002), an MBA from Wilfrid Laurier University with a concentration in innovation and entrepreneurship (2008), and a master's degree in management sciences with a concentration in technological innovation (2008).

Sarah Chawky is the innovation manager at Innovety, a regional innovation management consultancy firm specializing in digital and innovation management. She has extensive experience in working with teams from businesses of all sizes, to support them in capitalizing on technological and non-technological trends in order to develop their products, services, and process. Her experience extends to a number of sectors including telecommunication and information and communication technologies, engineering, oil and gas, and development. In 2018–2019, Sarah was the assistant regional officer at Terre des hommes (Tdh) for the MENA region's ICT for development (ICT4D), where she supported the region's four delegations to design new programs that are ICT-empowered, for increased scalability and sustainability. Sarah is a certified "Innovation Management Level 1: Innovation Associate" of the Global Innovation Management Institute.

Aya Abdel Halim is an assistant consultant at Innovety. She has contributed to multiple projects in the areas of innovation, entrepreneurship, and technology transfer and commercialization in various industries in Egypt and Lebanon. The projects she has worked on provided her with extensive knowledge on the innovative adoption of technologies for the creation of value, reaching of new markets, and streamlining of value chain processes. In addition to her work in innovation management, Aya holds a BA in economics from the American University in Cairo (2018) and was her class's recipient of the Ahmed Zewail Prize for Excellence in the Sciences and Humanities. She is also currently studying for an MA in economics at the American University in Cairo.

Agriculture

Abstract Agriculture is a wide global interest that encompasses numerous agricultural products such as food, fibers, fuels, and raw materials. Businesses that work in the agricultural industry participate in growing crops, holding farms, branching wood, and raising fish and animals (Vault 2020).

Keywords Farming · Produce improvement · Agricultural system · Agri-tech · Waste reduction · Optimum irrigation systems · Fully automated monitoring · Extended shelf-life · Solar drying · Eco-friendly · Genetically engineered · Full-journey tracking · Soil moisture detection · Biotechnology · Resource optimization · Crop optimization

Agriculture is a wide global interest that encompasses numerous agricultural products such as food, fibers, fuels, and raw materials. Businesses that work in the agricultural industry participate in growing crops, holding farms, branching wood, and raising fish and animals (Vault 2020).

The development of the agricultural industry has been affected by many factors including global warming, climate change, advancing technology, and human culture. Despite modern technological advances within the agricultural sector, farming methodologies are still heavily reliant on enhancing the environment and the quality of living organisms.

The industry continues to face a multitude of challenges that affect it at every juncture of its value chain, despite the positive attributes of technological advancements within it.

The very evolution of the production process which includes advancements in inbreeding and chemical and growth technologies has presented its threats to our environment and our overall human health (Vault 2020).

Farming professionals struggle to keep up with new enhancements in agricultural research, such as the competitive use of biotechnology to enhance livestock. These enhancements supply the market with healthier animals and advance breed development from other suppliers, which in turn produces more meat, eggs, and dairy products to fulfill consumer needs. Farmers who cannot utilize these technologies face delivering a lesser quality output (Vault 2020).

© Springer Nature Switzerland AG 2023
M. Anis et al., *Mapping Innovation*,
https://doi.org/10.1007/978-3-030-93627-3_1

Another challenge farmers face is that of global warming, which continues to affect them in numerous ways. Temperature changes can increase the rate of pest infestation and alter the levels of carbon dioxide found in the atmosphere (Rosenzweig and Hillel 1998). Pesticides and fertilizers have also been reported to be responsible for illnesses and pregnancy defects, and workers leave themselves prone to injury or even death when operating various machinery.

While the industry faces these challenges, the demand for food continues to increase driving it to struggle even more to meet demands.

The implementation of digital agricultural practices can support farmers in increasing their food production by allowing them easier access to data on crop cultivation. This in turn can increase their efficiency and productivity and reduce their operational costs.

Value Chain

The agricultural industry is mainly concerned with crops, livestock, and fisheries. The industry's value chain is comprised of six main components or phases: provision of inputs, growing and processing, packaging, storage distribution, end market, and waste management.

The first phase revolves around the selection and purchasing of "inputs" required for production. The next phase consists of farmers carrying out all activities needed for growing crops, raising livestock, harvesting fish, or creating produce. Such activities include technical maintenance of equipment and any farm facilities; properly maintaining the produce, livestock, or fisheries; and collecting the products. Afterward, the products need to go through a packaging phase to be appropriately stored for sales and distribution. Retailers and wholesalers can then sell the products for consumption during the end market phase. Wastes from the products themselves compose the last phase of this value chain, for when discarded they can and usually are reutilized for other purposes.

Case Studies

Case Study 1 LOCAL ROOTS

Vertical Farming

- Countries of implementation: USA, Canada, UAE, China, and South Korea
- Company: Local Roots Farms

Local Roots Farms is a business, founded in 2013, owned by Eric Ellestad, to improve global health by designing and operating high technology vertical farming tools. These could produce an improvement in the availability of food, lower negative impacts on the environment, and potentially provide better food at lower prices. Within a vertical farming agricultural system, growers can control variables such as light, humidity, and water to precisely measure year-round increases in food production with reliable harvests (Onica 2019).

Technology	Artificial intelligence (AR computer vision integrated with an artificial neural network		
Description	Local Roots Farms launched TerraFarm which is a vertical farming system in a 40-foot shipping container. The plants are grown stacked in several vertical layers. The technology implemented consists of computer vision with an artificial neural network that observes the plants remotely from California. The system uses 97% less water than normal as it collects evaporated water from air conditioning and reutilizes I (Vertical Farm 2020)		
Stage	In market		
Beneficiaries	Farm markets	• Better quality of food • Lower prices	
	Ministry of Agriculture	• Improved food quality and standards • Lowered cost • Potential of exportation (due to high quality) • Energy conservation • Less labor-intensive	

Case Study 2  Precision Decisions

Autonomous Harvest

Provision of Inputs → Growing and Processing → Packaging → Storage Distribution → End Market → Waste Management

- Country of implementation: UK
- Company: Precision Decisions

Precision Decisions is a company owned and founded by Clive Blacker. Originally a family business, the team expanded to improve farming services. Precision Decisions was acquired by Map Ag which is a company that specializes in agricultural analysis and designing technology to create an agri-tech data powerhouse that will guarantee a more valuable food supply chain for consumers and farmers (Mason 2018).

Technology	Artificial intelligence (AR) and automation robotics and autonomous vehicles	
Description	The Harper Adams University and Precision Decisions were able to create autonomous vehicles that harvested the first ever wheat crop to be produced by robots. The vehicles dispersed seeds and fertilized them, while agricultural machines such as tractors were utilized to harvest grain crops. As their technology evolved, they provided the machines with actuators, electronics, and robotic technology to control the machines remotely. The autonomous vehicle's destinations or stops are currently designed to be navigated by GPS, a feature that was predetermined by the researchers in the original design (Pultarova 2017)	
Stage	Testing	
Beneficiaries	Farmers	• Faster harvesting of crops • Less working hours • Less human error • Minimizing labor cost • Ease of use • Saving time
	Markets	• Increased food availability

Case Study 3

Agri Supply Chain

- Country of implementation: India
- Company: Gobasco

Gobasco is a company that works in the agri-tech, artificial intelligence (AR), and logistics industries. It was founded by Abhishek Sharma, Pranshu Aditya, and Vedant Katiyar in New Delhi, India, in 2017. The company aims to increase the supply food chain's yield and efficiency by using real-time data analytics and artificial intelligence (AR).

Technology	Artificial intelligence (AR)/machine learning	
Description	Gobasco tries to reduce wastage, improve prices and market discovery of crops, and raise farmers' accomplishment levels (Paul 2017). It uses artificial intelligence (AR)-optimized pipelines to increase the efficiency of the supply food chain. Computer vision and AI-based automatic sorting of vegetables and fruits provide standardized results that satisfy the industry benchmarks and allow farmers to grow their profits and create new opportunities for rural trade	
Stage	Testing phase	
Beneficiaries	Local farmers	• Increased profits • Increased trade nationally and internationally • Increased efficiency • Improved quality • Better resolution of agriculture maps • Competitive pricing • Lowered risks

Case Study 4

Precision Farming

- Country of implementation: USA
- Company: CropMetrics

CropMetrics is an agriculture company founded by Nick Emanuel in the year 2009. It aims to design and provide accurate technology solutions which increase water and nutrients while being energy efficient, encouraging natural resource conservation.

CropMetrics provides data-driven irrigation prediction to enhance guidance when it comes to crop, soil, and water development (CropMetrics 2018).

Technology	Internet of things (IoT)/variable rate irrigation (VRI)
Description	Variable rate irrigation (VRI) examines critical factors in fields and adjusts the rates of irrigation accordingly to ensure optimum levels of irrigation are met and maintained. Insufficient irrigation levels would result in a limited and premature yield while over-irrigation could produce lower yields and increased costs (CropMetrics 2018)
Stage	In market
Beneficiaries	Farmers

	• Improved yields
	• Enhanced water efficiency
	• Prevent unnecessary costs
	• Enhanced overall efficiency

Case Study 5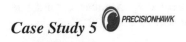

Agricultural Drones

- Countries of implementation: USA, Switzerland, New Zealand, China, Australia, and South Africa
- Company: PrecisionHawk

PrecisionHawk, founded by Christopher Dean and Ernest Earon in 2010, specializes in designing and integrating drone technology. It maps, tracks, and analyzes
farmlands using drones to improve efficiency, increase safety, manage resources,
and advance geospatial data analysis (PrecisionHawk 2020).

Technology	Internet of things (IoT)/agricultural drones (aerial mapping)	
Description	PrecisionHawk uses drones to gather valuable data. Farmers enter details of what field to survey and select an altitude or ground resolution. The drones collect multispectral, thermal, and visual imagery during the flight and then return to their launch site, providing fully automated monitoring of different farmlands (PrecisionHawk 2020). Data procured provides insight into crop health, numbers, estimated yield, average crop height, and numerous other key information	
Stage	In market	
Beneficiaries	Farming professionals	• Crop health imaging • Integrated GIS (geographic information system) mapping • Ease of use • Time-saving • Increased crop yield potential • Identifying optimal crop locations • Reduced health risks • Detailed data

Case Study 6

SYM Fresh Packaging

Provision of Inputs	Growing and Processing	Packaging	Storage Distribution	End Market	Waste Management

- Country of implementation: UK
- Company: Symphony Environmental Plc

Symphony Environmental Plc is a global company founded in 1995 with over 70 distributors worldwide. It develops biodegradable plastic solutions and designs technologies to improve nonbiodegradable plastic products (Symphony Environmental 2020).

Technology	Green technologies/biodegradable or protective technology
Description	SYM fresh bags are designed and produced with d2p ethylene absorber technology which absorbs and prevents ethylene from escaping. This process regulates moisture inside the bag, stemming dehydration and extending shelf life while maintaining the nutritional qualities and flavors of fruits and vegetables being preserved (Symphony Environmental 2020)
Stage	In market

Beneficiaries	Manufacturers	• It provides better protection against bacteria, fungi, mold, mildew, and algae for agriculture products • Increasing the shelf life of fruits and vegetables • Can be recycled • Waste management
	General public	• Less pollution • Reduces carbon emission

Case Study 7 KASCADE

Solar Drying

- Countries of implementation: Kenya, Uganda, and Mexico
- Company: Kascade

Kascade is a Dutch company that researches innovative solar solutions, using low tech to efficiently dry food products and produce clean water (Kascade 2020).

Technology	Green technologies/solar absorption	
Description	Kascade utilizes nature-dependent solar drying methodology to dry food products and maintain an ethos of being eco-friendly	
Stage	In market	
Beneficiaries	Farmer professionals	• Reduced costs • Increased convenience • Environmentally friendly • Minimized spoilage • Improved quality

Case Study 8

Waste Recycling

| Provision of Inputs | Growing and Processing | Packaging | Storage Distribution | End Market | Waste Management |

- Country of implementation: USA
- Company: Revolution

Revolution is a global company in Little Rock, Arkansas, that creates high-quality plastics while honoring its mission to preserve the environment by minimizing its carbon footprint. The company uses a process dubbed the closed-loop system where it produces and recycles plastics (Revolution 2020).

Technology	Green technologies/recycling	
Description	Revolution maintains eco-friendly production processes by using innovative recycled plastics as well as recycling their plastics within farmlands for other uses (Revolution 2020)	
Stage	In market	
Beneficiaries	Farm professionals	• Reduced agricultural waste • Reduced costs

Case Study 9

Transgenics

- Countries of implementation: USA and Germany
- Company: Bayer AG

Bayer AG is a life science company, founded by Friedrich Bayer in 1863, in Leverkusen, Germany. It specializes in health care and agriculture with a focus on pharmaceuticals, consumer health, and crop sciences (Bayer AG 2020).

Technology	Biotechnology/transgenics	
Description	Transgenics involves the insertion of one strand of DNA into another organism's DNA to introduce new genes into the original organism. This addition of genes into an organism's genetic material creates a new generation with the desired traits. The DNA must be prepared and packaged in a test tube and then inserted into the new organism. An example of transgenics is the rainbow papaya, which is modified with a gene that gives it resistance to the papaya ringspot virus (Bayer AG 2020)	
Stage	In market	
Beneficiaries	Farmer professionals	• Insect resistant • Herbicide tolerant • Disease resistant • Temperature tolerant • Drought tolerant • Capacity to grow under stressful conditions

Case Study 10

Oral Vaccines for Cancer

- Country of implementation: USA
- Company: Bayer AG

Bayer AG is a life science company, founded by Friedrich Bayer in 1863 in Leverkusen, Germany, and specializes in health care and agriculture. The company focuses on pharmaceuticals, consumer health, and crop science. Its pharmaceutical arm focuses on prescription products and specialty drugs, its consumer health arm focuses on nonprescription products in different categories, such as dermatology, and the crop science division focuses on agricultural land in seeds, pest control, and finally crop protection (Bayer AG 2020).

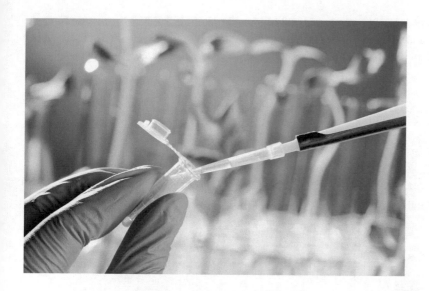

Technology	Biotechnology/genetically engineered crops	
Description	Genetically engineered crops are widely used in oral vaccines to trigger the immune system. Whether genetically engineered fruits or vegetables, this engineered product carries antigenic proteins which potentially boost patient immunity. Such vaccines have demonstrated great potential in the realm of cancer cures during preliminary studies (Phillips 2020)	
Stage	Testing phase	
Beneficiaries	Pharmaceutical companies	• Disease preventative vaccines
	Health-care systems	• Reduced patient health-care costs • Increased recovery rates

Case Study 11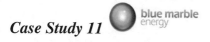

Biofuel

- Countries of implementation: USA, China, Germany, France, Canada, and Sweden
- Company: Blue Marble Energy

Blue Marble Energy was founded by James Stephens in 2005. It focuses on biotechnology and recycling to transfer waste biomass into renewable natural gas, using bacteria in the production process of biochemical and bioenergy products.

Technology	Biotechnology/genetically engineered crops	
Description	Genetic engineering and enzyme optimization techniques are used to develop better quality feedstocks and more efficient conversion and higher BTU outputs. High-yielding, energy-dense crops can minimize relative costs associated with harvesting and transportation, resulting in higher-value fuel products. However, the concept of using farmland to produce fuel instead of food comes with its challenges and solutions. The process relies on waste or other feedstocks that have not yet been able to compete on price and scale with conventional fuels (Amanda 2020)	
Stage	Testing phase	
Beneficiaries	The general public	• The clean and renewable energy source • Lowered pollution and emission levels • Reduced greenhouse gases • Reduced fuel costs

Case Study 12

Agricultural Trade

- Countries of implementation: USA and China
- Company: Louis Dreyfus Company

Louis Dreyfus Company, founded by Léopold Louis-Dreyfus in 1851, produces and transports nearly 80 million tons of products every year for nearly 500 million people. They aim to provide a supply chain from farmers to end consumers directly (LDC 2020).

Technology	Blockchain/blockchain-powered agriculture trade
Description	Blockchain technology is utilized for quality control in consumer product purchasing. Consumers have access to QR codes, found on the packaging, to track and follow the entire journey of any product they purchase allowing them to evaluate how the product was produced and handled (LDC 2020)
Stage	In market
Beneficiaries	Consumers

	Consumers	• Detailed tracking of food product • Real-time, supply, and demand information • Increased brand trust
	Agri-commerce participants	• Lowered costs • Faster payment options

Case Study 13 **MOTHIVE**

Mothive Ladybird

- Country of implementation: England
- Company: Mothive

Mothive is a tech company founded by Angelo Monteiro in 2015 to design and develop technologies that help farmers to maximize their efficiency, reduce waste, and improve the predictability of their crops (Mothive 2018).

Technology	Internet of things (IoT)/sensors and cloud-based software
Description	Mothive's Ladybird device contains sensors that detect air temperature, soil temperatures, humidity, light, and soil moisture every 8 minutes. It can predict crop diseases and expected growth, using the data collected, using sensors to determine favorable planting periods It also provides real-time recommendations, using cloud-based software via SMS or email, alerts on potential harmful scenarios, and automates irrigation and ventilation to optimize water usage in growing plants (Mothive 2018)
Stage	In market
Beneficiaries	Farmers

Beneficiaries:
- Reduced input costs
- Optimized water usage
- Improved yield
- Reduced labor costs
- Automated irrigation and ventilation

◻ ▪ BASF

Case Study 14 We create chemistry

Xarvio Scouting

- Countries of implementation: Global
- Company: BASF

BASF is a chemicals company founded by Friedrich Engelhorn in Ludwigshafen, Germany, in 1865. It specializes in chemicals, materials, agricultural solutions, surface technologies, industrial solutions, nutrition, and care (BASF 2020).

Technology	Artificial intelligence (AR)/machine learning	
Description	Xarvio Scouting is an application that spots disease in farmlands. The application alerts farmers to incidences of leaf damage and monitors for weed infestation. It offers farmers multiple tools to enhance crop growth economically and sustainably (Reiss 2019)	
Stage	In market	
Beneficiaries	Farmers	• Immediate weed detection • Plant disease detection • Insecticide spraying allotment and scheduling • Field monitoring • Estimation tools of nitrogen uptake in crops • Field risk assessments • Optimized crop production

Case Study 15

Camelina

- Country of implementation: Canada
- Company: Smart Earth Seeds

Smart Earth Seeds is a plant breeding company dedicated to producing the best Camelina plant, with maximum yield, while optimizing resources. It was founded by Jack Grushcow and focuses on the Camelina plant as it yields a high-quality oil, rich in nutrients (Smart Earth Camelina 2020).

Technology	Biotechnology/breeding	
Description	Camelina uses biotechnology to breed the Camelina plant using effective drought tolerance to produce nutrient-rich meals and oils. The processes used focus on seed yield, seed oil content, seed size, disease resistance, and herbicide tolerance. Their product is offered to businesses that use the plant for oil extraction and seed sales (Smart Earth Camelina 2020)	
Stage	In market	
Beneficiaries	Farmers	• Disease resistance • Herbicide tolerance • Drought tolerance • High yield • Low input costs • Production risk reduction • Agronomic advantages
	Markets	• Highly demanded products • Nutrition and dairy-rich merchandise

Case Study 16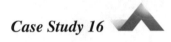

Metabolomics

- Country of implementation: USA
- Company: Metabolon

Metabolon specializes in a variety of fields, including drug discovery, health data, precision medicine, biochemical profiling, and many others. It was founded in 2000 and is currently expanding into tailored research and product development for interested clients.

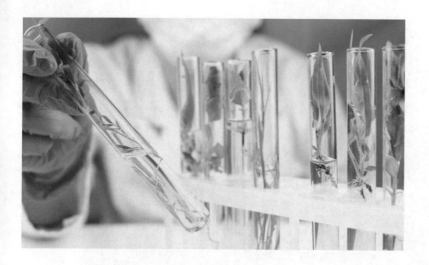

Technology	Biotechnology/biochemical profiling	
Description	Metabolomics is biochemical profiling that provides changes in plant tissues by identifying and monitoring plant development, plant nutrition, and biotic stress factors By observing and monitoring a plant, growers can optimize its resources and nutrition and make use of germplasm diversity (the mechanism through which living tissue can grow new plants (Metabolon 2020)	
Stage	In market	
Beneficiaries	Growers	• Identify plant diseases • Identify biotic stresses • Increased plant nutrition • Germplasm diversity • Identify abiotic stress factors

Case Study 17 BENSON◉HILL

CropOS™

- Country of implementation: USA
- Company: Benson Hill Biosystems

Benson Hill aims to improve crop production using design platforms that help enhance food systems and provide richer and healthier crops. Founded by Matthew B. Crisp and Todd Mockler in 2012, it specializes in artificial intelligence (AR), cloud biology gene editing among other things.

Technology	Biotechnology/genome editing and big data analytics
Description	The purpose of CropOS is to provide growers and consumers with better products by allowing researchers to control and predict outcomes by selecting and controlling wanted traits. Desired traits can include pest resistance, taste, and flavoring. Their products aim to facilitate research analysis and observation of seed growing data to increase the likelihood of producing seeds with the desired traits. Further analysis is then conducted to select seeds to be used for copying the superior genomes to other seeds (Benson Hill 2020)
Stage	In market
Beneficiaries	Growers and markets

Beneficiaries	Growers and markets

Beneficiaries | Growers and markets
- Healthier products
- Nutrient-rich products
- Crop optimization
- Disease-tolerant crops
- Insect-tolerant crops

Automotive

Abstract The automotive industry is an extremely competitive market due to progressive technological advances and product development: all essential for long-term growth and brand recognition.

Keywords Self-driving · Battery-electric vehicles · Car sharing · Additive manufacturing · Door-to-door navigation · Real-time tracking · Smart glasses · Supply chain transparency · Peer-to-peer network · Cloud-based platform · Fully autonomous vehicles · Bio-rubber tires · Net-zero energy · Biometric connection · Predictive maintenance · Autonomous emergency braking · Anomaly detection · Predictive collision alerts

The automotive industry is an extremely competitive market due to progressive technological advances and product development: all essential for long-term growth and brand recognition.

Research and development are core elements in the industry's value-added chain, enabling it to remain competitive in the marketplace and continue to offer a diversified product range and allow it to create millions of employment opportunities globally.

The main criteria that differentiate car brands are quality, design, technology, andperformance. However, there are many barriers to competing in this field given that it is a capital-intensive industry and, as such, tends to have elevated levels of operating leverage, needing extremely high-tech machinery, equipment, and high liquidity to operate (Pratap 2018).

Currently, the automotive industry is facing several challenges forcing manufacturers to adjust business models to meet market demands. These include, but are not limited to, environmental trends where customers are becoming more aware of their carbon footprints and so favoring ecologically friendly, more energy-efficient models. This has made many businesses feel the pressure of having to produce and develop electric vehicles which produce 50% less carbon dioxide emissions than their gas-powered.

competitors. Electric cars must now be offered to capture market opportunities and satisfy market needs.

Electric cars have slight differences in their selling and distribution chains as they require complementary products and services to be offered, such as the temporary battery chargers needed to complete a specific range of kilometers. These cars also need different charging stations, which, unlike gas stations, are not readily available everywhere (Burns 2020).

Technology is one of the leading challenges for manufacturers as continuous modifications in production assembly are essential while remaining practically rewarding, as prices increase in direct proportion to any updates being offered, e.g., the fully automated and self-drive vehicles available in today's marketplace. Technology also heavily impacts production lines, with manufacturers moving toward more modern, smart factories where robots provide a more efficient assembly and better-quality assurance (Automotive Alliance).

The current COVID-19 pandemic has resulted in automotive companies realizing the weakness in having a globalized supply chain. Industries are now beginning to develop more regionalized supply chains to support them in adopting and recovering more rapidly from any potential future disasters (Piparsania 2020).

Value Chain

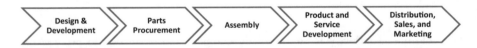

The automotive industry's value chain can be divided into five main phases: design and development, parts procurement, assembly, product and service development, and distribution, sales, and marketing.

The first phase of the value chain is comprised of researchers, automakers, and other players that are involved in research and development to design vehicles and modify their specifications.

The second phase of the value chain, parts procurement, includes activities that provide manufacturers with access to parts required for production. Relevant activities include the identification, selection, negotiation, and purchasing of the parts. These parts are then assembled by the manufacturers to produce functional products.

The product and service development phase is involved with increasing vehicle values by providing customer service, repairs, upgrade options, etc.

The last phase, distribution, sales, and marketing, then delivers the goods and services to their final consumer.

Case Studies

Case Study 1 TOYOTA

Self-Driving Vehicles

- Country of implementation: Japan
- Company: Toyota Motor Corporation

Toyota is an automotive company, founded in 1937; it produces and designs vehicles and sells them to more than 170 counties. It manufactures diverse types of vehicles, such as sports vehicles, commercial vehicles, etc. It is the second-biggest automaker in the world.

Technology	Automation/fully autonomous vehicles	
Description	Toyota's fully automated vehicles are called "e-Palettes" cars and are self-driving transparent boxes roaming around cities for delivering people, packages, and pizza. They are described as battery-electric vehicles that can be designed and customized to offer new features in mobile business services. The vehicle varies in size depending on its desired function (Hawkins 2018)	
Stage	Testing	
Beneficiaries	Businesses	• Pop-up stores • Unique consumer experiences • Doorstep services, e.g., delivery
	Health care	• Extended neighborhood reach • Quarantine facilities

Case Study 2 🔯 Reach Now

Car Sharing

- Country of implementation: USA
- Company: Reach Now

Reach Now is a company that produces mobility services allowing people to easily reach their loved ones. It was founded in 2016 and was later acquired by BMW.

Technology	Platforms/mobility	
Description	It is a mobile app offering car sharing services allowing users to either drive or hail rides to required destinations, with many offering features that enrich the sharing experience. It allows users to rent cars, or to rent out their cars, to earn a return on their investment. Unlike other platforms, BMW offers a fleet of Mini Cooper and BMW cars with the possibility of renting it for several days (Automotive UX 2016)	
Stage	In market	
Beneficiaries	Customers	• Easy car rental • Multi-preference application • Convenient payment methods
	Car owners	• Additional income potential

Case Study 3  **Betatype.**

3D Printing for Automotive Applications

Design & Development	Parts Procurment	Assembly	Product /Service development	Distribution, Sales, and Marketing

- Countries of implementation: Europe
- Company: Beta type

Beta type is a 3D printing company that offers automakers functional, 3D-printed components through their data processing platform that enables 3D printing with control over material, shape, and structure (Beta type).

Technology	3D printing/car components, 3D scanning, laser powder bed fusion
Description	The beta type can reduce parts cost production, from $40 per part to only $4, and decrease lead times from 444 hours to only 34 Beta type uses additive manufacturing to produce greater design freedom with minimal material by using laser powder bed fusion to produce and stack multiple parts on top of each other while reducing thermal stress and manufacturing time
Stage	In market

Beneficiaries	Automakers	• Significantly reduces waste • Cost-effective • Reduced manufacturing time • Increased efficiency • Reduced inventory stocking • Environmentally friendly

Case Study 4

BMW Connected Drive Application

Design & Development > Parts Procurment > Assembly > Product /Service development > Distribution, Sales, and Marketing

- Country of implementation: Germany
- Company: BMW

BMW (Bayerische Motoren Werke) is one of the top car manufacturers in the world and was founded in 1916. It a is German manufacturer with more than 100 years' experience in producing automobiles and motorcycles. Among other competitors, BMW is well-known for investing heavily in car technology and they are leaders in automotive infotainment systems.

Technology	Internet of things (IoT)/remote car monitoring and controlling	
Description	The BMW ConnectedDrive application is a mobile app that connects drivers to BMW cars remotely, monitoring and controlling many functions including fuel and fluid levels, GPS location, mileage, battery voltage, and emergency and when breakdown assistance may be needed. Using this application, users can estimate real-time arrival times, have remote access to essential functions of their BMW, send arrival information to friends and family, and have access to door-to-door navigation, including guiding users from their parking spots to their final destination	
Stage	In market	
Beneficiaries	Customers	• Interactive and engaging digital applications • New, embedded technologies • Accessible channels for more convenience and connectivity

Case Study 5

Augmented Reality (AR) HoloLens

- Countries of implementation: Sweden and Japan
- Company: Microsoft Corporation

Microsoft Corporation is a software company that offers, designs, and develops software products and services. It was founded by Bill Gates and Paul Allen in 1975. The company also provides gaming hardware segments including Xbox gaming and other entertainment accessories.

Technology	Augmented reality (AR)/AR glasses	
Description	Microsoft HoloLens, known as Project Baraboo, is a pair of mixed reality smart glasses developed and manufactured by Microsoft. HoloLens was the first head-mounted display running the Windows Mixed Reality platform under the Windows 10 computer operating system. HoloLens is integrated with many car assembly factories for more efficiency via visual digitalization. It enables production line workers to digitally view assembly instructions in real time. The HoloLens headset is used by manufacturers to support the design of cars, trucks, and SUVs without first having to build a prototype (Warren 2017)	
Stage	In market	
Beneficiaries	Manufacturers	• Increased efficiency • Increased consistency of production • High build quality • Increased detailing • Eliminated need for prebuilding

Case Study 6 ParkMobile

Self-Park and Pay

- Country of implementation: USA
- Company: ParkMobile, LLC

ParkMobile is the leading company in creating innovative ways to enhance smart parking and provide mobility solutions; it was founded in 2008.

Technology	Big data/navigation and autonomous services	
Description	ParkMobile is an app, connected to a platform, that helps navigate streets and finds parking slots with no interaction from drivers. Through ParkMobile, cars navigate and autonomously find parking slots and self-pay	
Stage	In market	
Beneficiaries	Customers	• Customer convenience • Mobile payment service • Virtual modification of set parking periods

Case Study 7 **MARELLI**

Transparency in Supply Chain

- Country of implementation: Japan
- Company: Marelli Corporation

Marelli is an automotive company that supplies and designs different products, including thermal systems, electronic products, and exhaust systems. It was founded in 2019 by Calsonic Kansei and Magneti Marelli.

Technology	Blockchain/cloud technology	
Description	Automotive supply chains are overly complex with many different industries involved, depending on the needs of the parts and the availability of raw materials, making it exceedingly difficult to be able to track the component's origin route. PartChain is a combination of blockchain and cloud technology which creates transparency in the supply chain, enabling the traceability of all components. Blockchain technology creates a peer-to-peer network, connecting suppliers and carmakers around the world. Using blockchain technology, companies like BMW know the logistics of raw materials or parts (Marelli)	
Stage	Recent to market	
Beneficiaries	Manufacturing companies	• Efficient management • Tracing of supply chain processes • Purchasing transparency

Case Study 8 automotiveMastermind®

Behavioral Analytics and Marketing Automation: Market EyeQ

| Design & Development | Parts Procurment | Assembly | Product /Service development | Distribution, Sales, and Marketing |

- Country of implementation: USA
- Company: Automotive Mastermind

Automotive Mastermind was founded in 2012 and is considered to be a leading provider of predictive analytics and marketing automation technology.

Technology	Big data and advanced analytics/platforms	
Description	Automotive Mastermind® technology captures data gathered from car dealers and merges them with big data. Gathered data includes information about customers, products, sociodemographics, specific customer targeting, and personalized marketing campaigns. The company's cloud-based platform helps dealers make precise predictions on automobile buying trends and automates the creation of microtargeted consumer communications, leading to proven higher sales and more consistent customer retention. Automotive Mastermind® currently works with more than 800 dealer partners, providing access to their technology to more than 6000 dealerships nationwide (Hunt 2019)	
Stage	In market	
Beneficiaries	Automotive manufacturers	• Increased sales • Improved customer retention • Bespoke products • Full market profile available
	Consumers	• Increased satisfaction • Enhanced customer experience

Case Study 9

Self-Driving Pod Waymo

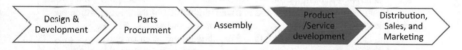

- Country of implementation: USA
- Company: Google

Google is a multinational company that produces and designs internet-related services and products. It was founded in 1998 by Larry Page, Sergey Brin, and Wesley Chan. Google's product portfolio includes Google Search, Knowledge Graph, Google Now, Product Listing Ads, AdSense, Google Display, and DoubleClick Ad Exchange and YouTube.

Technology	Automation/fully autonomous vehicles	
Description	Google has been working on self-driving technology since 2009 and currently has over ten million miles of stored data. Waymo can recognize many road features through sensors and software embedded in the system which are constantly scanning for objects or obstacles around the vehicle. In October 2019, Waymo started to invite members of its early rider program to take driverless rides with no humanoid operator behind the wheels (Waymo)	
Stage	Testing	
Beneficiaries	Users	• Reduced human risk error factors • Increased potential safety • A new era in mobility consumer reassurance

Case Study 10 COOPERTIRES

Guayule Plant-Based Rubber for Tire

- Countries of implementation: USA and Mexico
- Company: Cooper Tire & Rubber Company

Cooper Tire & Rubber Company is considered one of the leading competitors in the tire industry around the globe. It was founded in 1914, and it aims to produce and design high-quality tires that enhance its performance and can endure for thousands of miles.

Technology	Green technology/biotechnology	
Description	Cooper Tire & Rubber Company has currently introduced a tire design made from a shrub plant called guayule, found in Mexico and the USA, to put an end to the production and use of synthetic rubber in automobiles. This eco-friendly product, which is made of bio-rubber replacing synthetic rubber, was tested on different roads and tracks and found to have no difference in performances in comparison to traditional tires (Hevea) and bio-rubber tires (Sustainable Brands 2015)	
Stage	Recent to market	
Beneficiaries	Automotive industry	• Environmentally friendly • Cost-effective • Innovative product line
	Environment	• Environmentally friendly

Case Study 11 TESLA

The Giga Factory

| Design & Development | Parts Procurment | Assembly | Product /Service development | Distribution, Sales, and Marketing |

- Country of implementation: USA
- Company: Tesla

Tesla is an automotive company founded in 2003. It aims to increase the use of sustainable energy by increasing the production of affordable electric vehicles and renewable energy production and storage. Tesla ensures reliability, performance, and safety in its vehicles.

Technology	Green technology/solar energy
Description	Giga Factory is housed on 1.9 million square feet, with 5.3 million square feet of that given to operational space across several floors. It is being constructed in various stages to allow tesla to begin production at once while continuing the expansion of its premises. The factory is expected to be the biggest building in the world and designed to have net-zero energy, being powered primarily by renewable energy sources (tesla)
Stage	Recent to market (partially working and in development for further expansion)

Beneficiaries	Environment	• Cost-effective • Better public accessibility • Eco-friendly • Reduced waste

Case Study 12

Avatar Concept Car

Design & Development	Parts Procurment	Assembly	Product /Service development	Distribution, Sales, and Marketing

- Country of implementation: Germany
- Company: Mercedes-Benz

Mercedes-Benz is one of the largest sellers of premium vehicles in the world. It was founded in 1926 by Karl Benz, Gottlieb Daimler, Wilhelm Maybach, and Emil Jellinek and aims to produce luxury vehicles that provide customers with comfort and safety.

Technology	Biometrics/green technology	
Description	The vehicle forms a biometric connection with drivers. The conventional steering wheel is replaced with a control unit embedded in the central console which monitors drivers based on their heartbeat and breathing upon contact. The vehicle is also eco-friendly, aligned with avatar themes and the cars' battery is made with composite materials and fully recyclable, so aligns with global efforts to reduce reliance on fossil fuels (Mercedes Benz 2020)	
Stage	Idea stage	
Beneficiaries	Customers	• Improving and personalizing the customer experience through biometrics • Can identify the driver based on heartbeat and breathing, therefore, reduces chances of stealing • Environmentally friendly
	Environment	• Aligned with global efforts to reduce reliance on fossil fuels

Case Study 13 SIEMENS

MindSphere

- Country of implementation: Germany
- Company: Siemens

Siemens is an electronics engineering company, founded by Johann Halske and Werner Siemens in 1847. It specializes in various sectors, including health care, energy, and industrial. Siemens provides innovative IT services and solutions, insurance solutions, financial products, and others.

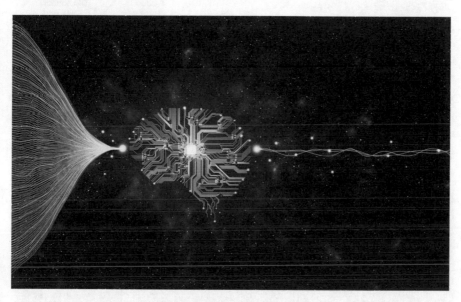

Technology	Internet of things (IoT)/artificial intelligence, data analytics, and sensors
Description	MindSphere is a cloud-based operating system that allows other technologies or machines to collect data. This will help manufacturers observe the industrial assets using machine learning technology. MindSphere can connect all your machines, technologies, and systems which enables users to use the valuables of the data using advanced analytics. Sensors are also being used to collect information from different machines and then download them onto the industry's database in the cloud operating system. This collected information is then uploaded to the machine learning algorithm of the system to see these analytics and upload them onto a dashboard. Moreover, MindSphere can detect problems in the machines that are underperforming and define them. It then informs the maintenance operators about the type of problem exactly and recommends they either try to fix it or completely shut it down. MindSphere is also able to connect various applications (Jesus 2018)
Stage	In market
Beneficiaries	Industries

Beneficiaries	Industries	• Monitored data in real time • Prediction of performance challenges • Improves efficiency and profitability • Connected applications to increase data significance

Case Study 14

Robust Sensing for Autonomous Driving

| Design & Development > | Parts Procurment > | Assembly > | Product /Service development > | Distribution, Sales, and Marketing > |

- Countries of implementation: Japan
- Company: Mitsubishi Electric Corporation

Mitsubishi Electric Corporation is a company that designs, develops, and sells electrical machines that can be used in different systems; it was founded in 1921. It designs equipment for different sectors including energy and electric systems, communication systems, industrial automation, and others (Crunchbase).

Technology	Automation/artificial intelligence
Description	Mitsubishi Electric developed a sensing technology for the detection of the car's perimeters even in the presence of fog or rain, such as the positions, velocities, and sizes of any obstacles or vehicles surrounding it. This will enable the vehicle to be able to drive autonomously even in rough weather. This newly developed system can collect data from various sensors including time series data which collects information on the width, orientation, distance, velocity, etc. of the surrounding vehicles from sensors in real time. The technology is designed also using autonomous emergency braking (AEB) to test the performance of the vehicles in bad weather. Autonomous emergency braking (AEB) can safely perform emergency braking even if the sensors are not able to perform well. The LiDar embedded in this system can detect and monitor the surrounding environment using pulse laser signals. The millimeter-wave radar can detect velocities and distances of surrounding vehicles, and the camera can recognize the obstacles' sizes (Mitsubishi electric)
Stage	Testing
Beneficiaries	Customers

		• High accurate detection is achieved in rough weather • Decrease the number of accidents in rain and fog conditions • High performance of autonomous braking system • The technology can detect the weather of the surrounding environment using LiDar • High accuracy detection of velocity, distance, sizes, etc.

Case Study 15

FreeMove

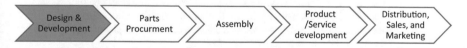

- Country of implementation: USA
- Company: Veo Robotics

Veo Robotics is a company that develops innovative solutions to improve the performance of collaborative robotics. It was founded by Clara Vu, Patrick Sobalvarro, and Scott Denenberg in 2016. It designs products that use computer vision, artificial intelligence, and 3D sensing.

Technology	Artificial intelligence/3D sensing and computer vision	
Description	FreeMove is a technology-designed system that can combine the strengths and precision of industrial robots while at the same time using the creativeness of humans; it allows both humans and robots to work in the same environment or workplace while at the same time ensuring safety. Veo FreeMove consists of three systems: FreeMove sensors, FreeMove engine, and FreeMove studio. The FreeMove sensors consist of 3D sensors that are located on the work cell of the machine. The FreeMove engine is a high-performance computing system that processes and reads data collected from the sensors and ensures that there is a safe separation distance kept between the robot and humans. Hence the FreeMove engine monitors the distance that can be traveled by the robot to bring it to a safe stop. The FreeMove studio is a software system used for real-time visualization of the FreeMove sensors and FreeMove engine. The FreeMove systems can position all objects in their environment by calculating the separation distance then signaling the robots to either completely stop or slow down for the human's safety (Veo robotics)	
Stage	Testing	
Beneficiaries	Industries and factories	• Able to combine both worker and robots in the same place with no health or safety risks • Increases productivity • Enhancing workers' efficiency • Reducing capital and operating expenses • Reduced downtime • Greater manufacturing flexibility
	Workers	• Increases the safety of workers • Enhancing their skills

Case Study 16  NHTSA

Vehicle-to-Vehicle Communication

```
Design &       Parts                      Product        Distribution,
Development    Procurment    Assembly     /Service       Sales, and
                                          development     Marketing
```

- Country of implementation: USA
- Company: National Highway Traffic Safety Administration (NHTSA)

NHTSA is founded by the Highway Safety Act in 1970; its goal is to try to find ways that can reduce crashes and accidents which in turn reduces the attendant costs. It tries to find ways to enhance safety and produce high standards in motor vehicles (Crunchbase).

Technology	Artificial intelligence/sensors
Description	Vehicle-to-vehicle communication allows vehicles to wireless communication with each other by exchanging data about their speed, direction, location, and movements. The technology slows vehicles to send and receive omnidirectional messages ten times per second which creates a 360-degree information about all surrounding vehicles. The vehicles can use this information to determine crash threats and hence this technology produces visual and audible alerts to warn drivers about any threats or obstacles. The information that is sent has a range of more than 300 m and can detect any danger, such as traffic or changes in the weather. This technology also has cameras and radars to monitor or detect any collision threats. This technology can be used by cars, trucks, buses, and motorcycles (NHTSA)
Stage	Testing

Beneficiaries	Users	• Increase the safety of drivers • Decrease crashes or any safety risks • Enhances the performance of vehicle's safety systems • Avoid traffic jams • Maintain a safe distance from other cars • Provides direction and route optimization • Assists by giving simple warnings

Case Study 17 ⌐ Progress

Predictive Analytics Software: DataRPM

Design & Development	Parts Procurment	Assembly	Product /Service development	Distribution, Sales, and Marketing

- Country of implementation: USA
- Company: Progress

Progress is a computer software company that was founded in 1981. It is the leading company in developing and designing strategic business applications. It aims to deliver high-quality digital experiences while minimizing the time and cost of partners and customers. It specializes in machine learning, application development, data connectivity, and others (LinkedIn).

Technology	Artificial intelligence/machine learning	
Description	DataRPM is considered to be a cognitive intelligent software or in other words an anomaly detection and prediction software that can use the data coming from the industrial internet to enable industries to improve product output and quality. Industries can use this software to identify external factors that could affect the production of manufacturing engines. Moreover, using DataRPM manufacturers can spot engine failures weeks before it occurs due to machine learning vision. The software collects past data, current data, and any external factor's data so that it can establish a minimum starting point for comparison to measure the performance of the engines on an ongoing basis. Also, this technology helps automotive manufacturers in collecting and analyzing all data and generate possible recommendations to increase and enhance efficiency (Progress)	
Stage	In market	
Beneficiaries	Automotive industries	• Suggests possible recommendations in increasing efficiency and productivity • Predicts automated failures which reduce malfunctions, maintenance costs, and breakdowns • Optimize inventories • Improvement in delivery time • Optimizes resources

Case Study 18

Robo-Glove

- Country of implementation: USA
- Company: General Motors

General Motors is an automotive company that has the goal of producing or designing technology to produce a safer and more sustainable environment. Moreover, its goal is to create a world with zero emissions (LinkedIn).

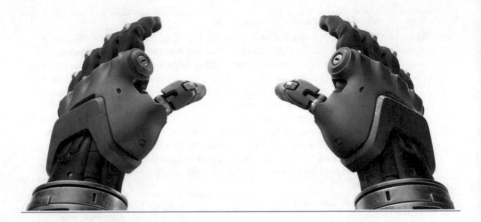

Technology	Artificial intelligence/robotics technology and sensors
Description	Robo-Glove allows humans to easily grasp heavy objects or any tools to reduce fatigue in hand muscles. It is designed using actuators that are embedded in the upper part of the glove to give support to the human fingers. It also has pressure sensors that are embedded in the fingertips of the Robo-Glove so that it can detect when the person is holding a tool. When the user is holding a tool, the synthetic tendons in the glove withdraw which pulls the fingers into a gripping position until the pressure sensor is released. The actuators and tendons embedded in the glove are used to mimic the muscles of the human hand. The Robo-Glove is powered by a lithium-ion power tool battery and a belt clip. This glove is a very lightweight technology that fits perfectly on the hand. It is also designed for multiple hands and arm sizes (Vincent 2016)
Stage	In market
Beneficiaries	Automotive repair workers	• Reduces muscle strain from doing repetitive tasks or carrying heavy objects • Lightweight technology; therefore, the user barely feels it • Increases productivity in demanding applications

Case Study 19 𝕟 nauto™

Predictive Collision Reacts

Design & Development	Parts Procurment	Assembly	Product /Service development	Distribution, Sales, and Marketing

- Country of implementation: USA
- Company: Nauto

Nauto is a computer software company that specializes in artificial intelligence to produce products or devices to predict and decrease high-risk events in the mobility ecosystem.

Technology	Artificial intelligence/biometrics and data analytics	
Description	Nauto developed a device that will be able to collect and analyze real-time data and alert the driver of any possible collisions. Predictive collision alerts reduce collision that can be caused by other vehicles, cyclists, changing lights, pedestrians, and others. It collects data about braking, speed, and movements of other vehicles from all angles. This technology is continuously scanning and analyzing the facial expressions and movements of the driver so that it can detect unsafe driver behavior. It gives alerts that are twice as loud to distracted drivers. There are three driver alerts: Mild, medium, and severe depending on how much the driver is distracted. If a vehicle is driving at 60 mph, the device gives alerts at 100 ft. This device keeps drivers attentive to reduce collisions and traffic violations (Nauto)	
Stage	In market	
Beneficiaries	Users	• Reduction in accidents and collisions • Increases the alerts depending on how much distracted the driver is • Reduce maintenance costs • Increase safety • Reduce traffic violations and their costs

Banking and Financial Services

Abstract The finance industry provides financial services to people and corporations through institutions such as banks, investment houses, and real estate brokers. As per the International Monetary Fund's finance and development department, financial services are all operations by which individuals and businesses earn economic goods; for instance, a payment provider who accepts and transfers funds between a payer and recipient is deemed to be offering a financial service (Investopedia 2019).

Keywords Technological inclusion · Transaction tracking · Peer-to-peer lending · Robot advisor · Revenue management · Risk management · Customer purchasing behavior · Fundraising · Long-term planning · Portfolio formulation · Online trading · AI-powered robots · Simulation · Anti-laundering · VR experience

The finance industry provides financial services to people and corporations through institutions such as banks, investment houses, and real estate brokers. As per the International Monetary Fund's finance and development department, financial services are all operations by which individuals and businesses earn economic goods; for instance, a payment provider who accepts and transfers funds between a payer and recipient is deemed to be offering a financial service (Investopedia 2019).

One of the significant challenges in the banking industry is regulatory compliance, resulting from the substantial increase in regulatory fees relative to earnings and losses since the 2008 global financial crisis. As there are harsh results for non-compliance, banks try to stay current with the latest regulatory changes, which incurs further costs for them. Also, today's generation of customers has become smarter and more aware, expecting a high level of personalization, convenience, and financial experience. Today's customer knows more about technology and expects a highly developed digital experience. This incurs a dilemma for the banking industry and credit unions who now need to satisfy multiple generations. Recently, security breaches have become significant challenges and require financial institutions to develop technology-driven security measures to keep their customer data as safe as possible (Wingard 2020).

   Competition within the banking and financial service providers has encouraged them to seek innovations aimed at creating unique customer experiences (Bharadwaj 2020), which has advanced technologies within the banking sector.

   COVID-19 has resulted in a state of emergency for the world's economy. Many substantial companies and entrepreneurs are facing an imminent crisis. Banks have traditionally been seen to be responsible for stabilizing economies and rescuing businesses and freelancers affected by unstable market factors, such as the corona pandemic. The banking industry is now faced with additional challenges in that customers wish to avoid handling banknotes in an attempt to prevent the risk of spreading the disease. This has led to many services now being provided online (Rehfish 2020).

Value Chain

The value chain for banking and financial services is comprised of five main phases: marketing activities that aim to develop and promote the services available, sales to provide revenue, product for continuous development financially or technically to increase service value, transaction that involves banking and financial services, and finally the end customer phase, which includes all after-sales and customer service activities.

Case Studies

Case Study 1

Data Analytics in Personal Financial Management

- Countries of implementation: South East Asia and Eastern Europe
- Company: Matchi

Matchi is an international platform that works on connecting financial institutions with individuals and other large incorporations. Matchi always provides leading-edge and innovative technology solutions.

Technology	Big data and advanced analytics	
Description	Thanks to big data and advanced analytics, there is more technological inclusion for individuals, banks, and stores. PFM solution has two main features. The first is recording and tracking transactions and aggregating this information through different products to provide insight on current finances. The second is to provide an automated solution for saving and classification of personal transactions and tracking of budgets (Matchi.biz 2020)	
Stage	In market	
Beneficiaries	Newlyweds	• Financial support • Access to home and utilities support payments
	Individuals	• Living or small projects finance • Help microfinance small investments

Case Study 2

Internet of Things (IoT) in Money Lending

- Country of implementation: USA
- Company: Peerform

Peerform is a platform for peer-to-peer lending. It connects people who need to borrow money with people interested in investing it. It provides up to 3-year terms of personal loans, ranging from 1000 USD to 25,000 USD.

Borrowers list the needed amount with a fixed rate that they can manage and lenders set their investing amounts and the required rate of return they would like.

Technology	Internet of things (IoT)
Description	Peer-to-peer lending is considered an alternative to the banking and financial services system in providing better interest and borrowing rates. It connects individuals with investors willing to fund personal loans (Anna 2020)
Stage	In market

Beneficiaries	Startups	• Financial support
	Microfinancing	• Financial on savings • Personal projects support

Case Study 3 Betterment

Robo-Advisory

- Continents of implementation: USA and Europe
- Company: Betterment

Betterment is one of the leading companies in the Robo-advisory industry. It uses technology to recommend portfolios and automate the investment process.

Technology	Artificial intelligence (AI)/machine learning	
Description	This tool provides retail banking customers with a trading advisor called a robot-advisor. This tracks and analyzes customers' income and portfolio funds, including expenditure. The customer then sets goals for investments and the website aggregates this to provide a suitable financial goal (Ayan 2019)	
Stage	In market	
Beneficiaries	Individuals	• Increased income • Financial support

Case Study 4

Electronic Currencies

- *Coun*try of implementation: USA
- Company: Circle

The Circle is a company that provides a stabilizing monetary value through something like a bitcoin called USDC (US Dollars Coin). USDC is considered a breakthrough in using money.

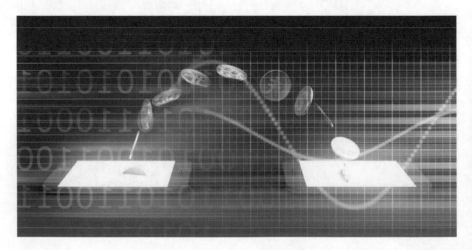

Technology	Internet of things (IoT) and automation
Description	This tool allows individuals and companies to purchase their digital dollar's value with their money to make fast online payments. This tool did not only provide payments for purchases but also money transfer in different currencies as it stabilizes this to the value of the digital dollar (Glassdoor 2020)
Stage	In market
Beneficiaries	Families

Beneficiaries Families	• Facilitated monetary transactions • Easy, fast, and affordable payment transfers

Case Study 5 FINASTRA

Banking Platforms

- Continents of implementation: North America and Southeast Asia
- Company: Finastra

Established in 2017, Finastra is a company that offers innovative technologies for corporate financial needs. Finastra's corporate banking platform provides corporate applications and developed APIs through cloud-based systems.

Technology	Platform	
Description	Finastra helps corporations map their business channels across business platforms. The application can be integrated with corporate systems to provide financial services such as revenue management, risk management of risk, and payment (Finextra 2020)	
Stage	In market	
Beneficiaries	Corporates/financial institutions	• Financial risks forecasts

Case Study 6

Tech Retail Banks

Marketing 〉 Sales 〉 Product 〉 Transaction 〉 Customer

- Countries of implementation: USA and Nepal
- Company: Arttha

It is a platform that is digitally built and considered to be one of the fastest-growing fintech solutions. They provide retail banks with digital solutions for their banking services such as mobile banking, microfinancing, insurance, and branchless banking.

Technology	Platform	
Description	Arttha is an application that can integrate with corporate systems to provide financial services such as revenue management, risk management, and payments, making a seamless experience using technology platforms (Arttha 2020)	
Stage	In market	
Beneficiaries	Retail banks	• Branchless banking • Online banking

Case Study 7

Platforms

Marketing ⟩ Sales ⟩ Product ⟩ **Transaction** ⟩ Customer

- Country of implementation: Egypt
- Company: WeAccept

A company founded in Egypt that provides financial services for individuals to help them make their payments.

Technology	Platform
Description	WeAccept platform helps businesses accept customer payments and track customers' purchasing behavior (WeAccept 2020)
Stage	In market
Beneficiaries	Businesses	• Affordable online payments
	Consumers	• Convenience

Case Study 8

Financial Support Made Available on Platforms

| Marketing | Sales | Product | Transaction | Customer |

- Country of implementation: USA
- Company: CircleUp

CircleUp provides startups and entrepreneurs with financing and encouragement to finance their projects.

Technology	Platform
Description	CircleUp helps entrepreneurs and small businesses raise funds to finance their projects. It helps people interested in investing small sums of money to increase their income; hence, it is considered as a funding and lending platform (CircleUp 2020)
Stage	In market
Beneficiaries	Startups/entrepreneurs · Finance for youth projects

Case Study 9

Digitalizing Financial Processes

Marketing | Sales | Product | Transaction | Customer

- Country of implementation: Egypt
- Company: WePay

WePay started offering financial solutions to the Egyptian market in collaboration with the Egyptian bank starting in 2019.

Technology	Automation and digitalization	
Description	We provide a mobile app that facilitates payments such as money transfer and utility bill payment (The economical parliament 2019)	
Stage	In market	
Beneficiaries	Families and youth	• Bill payment • Monetary transactions between family and friends

Case Study 10

Internet of Things (IoT) and Mastercard

- Country of implementation: Egypt
- Company: Meeza

Meeza is a domestic payment system to provide easier transactions for individuals and families which is provided by CIB banks.

Technology	Internet of things (IoT)	
Description	CIB offers a mobile app that provides and facilitates payments for individuals while supporting Egyptian government systems, as it is officially regulated by the Central Bank of Egypt (CIB,2019)	
Stage	In market	
Beneficiaries	Consumers	• Bill payment • Convenient financial transactions

KENDRIS

Case Study 11 PERSONAL | INDEPENDENT | DIGITAL

Websites Development in Management Services

- Country of implementation: Switzerland
- Company: Kendris

Kendris is a multifamily office that is based in Zurich, Switzerland. They offer asset management services for wealthy families.

Technology	Platform/website development		
Description	Kendris uses a website that provides its services through consultancy and direct investment, with long-term planning solutions to help old couples with retirement plans who would like to have the money that they have made with their hard work to serve them as a source of income and not to lose the value of their money and assets (Kendris 2020)		
Stage	In market		
Beneficiaries	Wealthy seniors	• Asset management • Income generation • Commission charges	

Case Study 12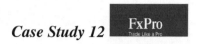

Platforms in Global Investments

- Country of implementation: Online global
- Company: FxPro

FxPro started first as a brokerage firm that invested mainly in metals. Afterward, they created their platform to provide investors globally to invest their money in any business and any location in the world.

Technology	Platform
Description	FxPro is a global platform providing investors with all the tools they need for their investment such as portfolio formulation and guides on how to trade online. The good thing is that it provides a 0% commission on funding (Peter 2020)
Stage	In market
Beneficiaries	Investors

- Faster results
- Accurate and precise decision-making

Case Study 13

Conversational Artificial Intelligence (AI)

Marketing	Sales	Product	Transaction	Customer

- Country of implementation: USA
- Company: Kasisto

Kasisto is one of the banking industry's leading digital institutions, founded to master the language of banking and financial services.

Technology	Artificial intelligence (AI)/machine learning	
Description	Kasisto contributed to conversational artificial intelligence (AI) by allowing banks to build their chatbots and virtual assistants, using AI reasoning and natural language understanding to handle sophisticated questions about financial management that other regular digital customer services cannot answer (Stephen 2019)	
Stage	In market	
Beneficiaries	Banking institutions	• Artificial intelligence (AI) assistance
	Customer	• Artificial intelligence (AI) bot guidance and support

Case Study 14 :) **Affectiva**

Smart Robots at HSBC

Marketing >> Sales >> Product >> Transaction >> **Customer**

- Country of implementation: USA
- Company: Affectiva

Affectiva specializes in emotional measurement technology, out of MIT's Media Lab. They developed software to recognize human feelings based on facial or physiological responses.

Technology	Artificial intelligence (AI)/software development	
Description	Pepper incorporated artificial intelligence (AI) and machine learning algorithms when the MIT media lab launched Affectiva. They developed it with intelligent abilities to sense emotion and cognitive states. Afterward, HSBC used this update on bank floors to handle hosting duties such as walking the customers through the opening of accounts, cracking jokes, and relaying credit card details (Stephen 2019)	
Stage	In market	
Beneficiaries	Customers	• Improved customer service • Improved customer support • Time efficiency
	Banking HR departments	• Cost-effective

Case Study 15  SIMUDYNE

An Investment Simulation Analysis

- Country of implementation: UK
- Company: Simudyne

Simudyne is a simulation analysis company that offers companies a new technology to apply agent-based modeling.

Technology	Artificial intelligence (AI)/machine learning	
Description	Simudyne is a technology provider that offers agent-based modelling and machine learning to test and simulate millions of market scenarios. Simudyne's technology allows financial institutions to test different stress analyses and simulations for market contagion, on large scales. The company's CEO Justin Lyon announced in the Financial Times that the simulation helps investment bankers spot tail risks and low-probability high-impact events (Stephen 2019)	
Stage	In market	
Beneficiaries	Financial institutions	• Perform market testing • Predict market testing outcomes • Provide data • Supports informed financial choices

Case Study 16 **AYASDI**

Machine Learning in Fraud Detection

Marketing → Sales → Product → Transaction → Customer

- Country of implementation: USA
- Company: Ayasdi

Ayasdi is a software company that uses software platforms and applications to build different models which predict possible outcomes using big data and high-dimensional data sets.

Technology	Artificial intelligence (AI)/machine learning	
Description	The volume and complexity of anti-laundering are considered very difficult for humans. Ayasdi supports this by using three main methods to incorporate machine learning in the process. It first uses intelligent segmentation to produce the fewest number of false positives. It then uses an advanced alert system that auto-categorizes alert priorities. It finally uses advanced transaction monitoring that exploits machine learning to spot suspicious anomalies (Stephen 2019)	
Stage	In market	
Beneficiaries	Taxpayers	• Diminishes money laundering • Protects taxpayers

Case Study 17

Virtual Reality (VR) Banking Apps

- Country of implementation: France
- Company: NP Paribas

NP Paribas is a global financial services company that offers solutions in capital markets and securities services.

Technology	Virtual reality (VR)	
Description	BNP Paribas developed a virtual reality (VR) application that allows retail bankers to monitor financial activities and transactions through a VR environment. Furthermore, they have developed a separate application that walks the user through the steps involved in buying a new home (Jeffry 2020)	
Stage	In market	
Beneficiaries	Bank users	• Simplified financial data

Case Study 18

VR in Citi Banks

Marketing > Sales > Product > Transaction > Customer

- Country of implementation: USA
- Company: Citi

Citi banks offer financial services such as cash management, transaction services, and capital market operations.

Technology	Virtual reality (VR)	
Description	Citi has launched a series of live virtual reality (VR) concerts to attract more users to join them. They named it "Backstage with Citi" as it will reward selected card members with special events. Fans responded positively by enjoying these live shows and virtual "backstage experiences" which would stream with popular artists using VR headsets (Jeffry 2020)	
Stage	In market	
Beneficiaries	Citi	• Increased customer loyalty
	Customers	• Exclusive membership events and promotions

Construction

Abstract The construction sector is one of the largest world economy sectors and employs around 7% of the workers of the world (McKinsey Global Institute 2017). It focuses on jobs related to demolition, renovation, or maintenance of any infrastructure.

Keywords 3D design · Low-cost material · Virtual touring · Natural lighting sources · Concrete monitoring · Automatic navigation · AR visualization · Equipment rental · Connected maintenance service · Humanoids robotics · Tracking and recording systems · 5D building · Integrated platforms · Smart sensors

The construction sector is one of the largest world economy sectors and employs around 7% of the workers of the world (McKinsey Global Institute 2017). It focuses on jobs related to demolition, renovation, or maintenance of any infrastructure.

About $ 10 trillion is spent on construction goods and services annually, which accounts for 13% of the world's GDP. The industry's functionality ranges from planning to final structural services; hence, analyzing challenges facing the sector is essential.

Egypt's large, rapidly expanding population is one of its greatest economic assets. It is the world's 14th most populous country, Africa's third most populous nation, and the most populous in the Arab world. According to the World Bank, this population is growing by 2% per year (Oxford Business Group 2019). This creates an increasing demand for construction which is challenged by a labor shortage (Conti 2019).

Construction has been impacted by the COVID 19 pandemic as it requires intensive labor and depends heavily on a supply chain for both its labor and materials, critical components of all construction projects (European International Contractors 2020).

Technological solutions provide a great solution as they can reduce the workforce, increase productivity, and execute tasks remotely. Utilizing big data, IoT, and VR can help designers and developers visualize projects, eliminate uncertainties, and provide their consumers with a more rewarding overall experience.

Value Chain

The construction sector's value chain is comprised of seven phases: material sourcing, material processing, design, construction, sales and marketing, occupation/observations and measurements, and waste management.

In the first two phases, industry players are involved with materials used in construction. Construction companies purchase, negotiate, and select materials needed. Once project designs are specified and finalized, they are shared to be executed in the third or construction phase. Sales and marketing activities are carried out to promote projects and streamline sales. Finally, debris and waste from construction sites, which may include packaging, unused materials, or demolition materials, are then discarded or reused in the final phase.

Case Studies

Case Study 1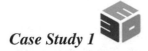

3D Printing in Construction

- Country of implementation: Egypt
- Company: Etba3ly 3D

Etba3ly is an Egyptian venture based in Cairo that works in the field of fused deposition modeling (FDM) 3D printing. They offer the design, printing, and delivery of 3D models.

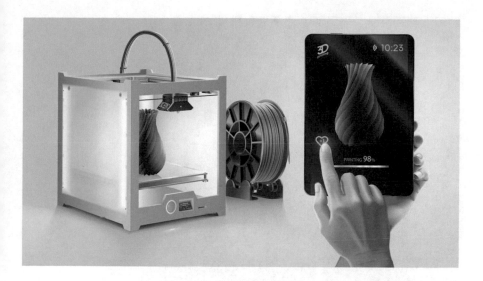

Technology	3D printing/FDM 3D printing	
Description	Etba3ly 3D is a platform specializing in 3D printing covering the entire cycle from design to delivery. Etba3ly allows its users to either create their designs online through a free online tool or connects them to local designers. Next, Etba3ly 3D prints these designs and delivers them to users across Egypt Etba3ly 3D produces a wide range of products, such as architecture and interior design 3D models, consumer product prototypes, translucent decorative elements, and items created with materials resistant to salts, acids, alkalis, and solvents (3D Printing Egypt – On-Demand 3D printing in Egypt)	
Stage	In market	
Beneficiaries	Designers, real estate developers, students, and entrepreneurs	• Time-effective • Cost-saving

Case Study 2 ꒱ apis cor

3D Printed Houses

- Countries of implementation: Global
- Company: Apis Cor

Apis Cor is a construction company headquartered in Russia that specializes in the building of robotically constructed houses and the use of 3D printing technology.

Technology	3D printing	
Description	Apis Cor is one of the industry pioneers in 3D printing construction. They have created their 3D printer that covers an area of 132 m^2. Their technology allows them to build in any location within hours with the help of a mobile feature	
Stage	In market	
Beneficiaries	Developers, customers	• Shortened construction time • Precise execution of architectural molds • Cost savings • Location friendly

Case Study 3

Virtual Reality Tours

Sourcing › Material Processing › Design › Construction › Sales & Marketing › Occupation/ O&M › Waste Management

- Countries of implementation: Global
- Company: Klapty

Klapty is a Switzerland-based startup specializing in 360 virtual tours.

Technology	Virtual reality/360 videos
Description	Klapty is a free premium-based platform that enables individuals and businesses to create, edit, and circulate 360 virtual tours. The platform targets realtors, photographers, hotel owners, companies, and individuals wishing to share content visually. Klapty is compatible with all 360 cameras and images (JPG). The platform acts as a marketplace for 360 content creation and can connect users. Members can upload panoramic photos and use different features to enhance their material
Stage	In market
Beneficiaries	General public

Beneficiaries second column:
- Cost-effective marketing and sales
- Accessibility and scalability
- Enhanced customer experience

Case Study 4

Green Construction Material

- Country of implementation: Stockholm
- Company: KTH Royal Institute of Technology

KTH Royal Institute of Technology is one of Sweden's largest technical research universities. It is one of the country's greatest centers of innovation and research, especially in the fields of technology and engineering ("KTH Royal Institute of Technology").

Technology	Green/eco-friendly material	
Description	Swedish researchers at KTH Royal Institute of Technology in Stockholm have developed an 85% transparent alternative material to wood by compressing strips of wood veneer and replacing lignin with polymer through a process that is similar to pulping. This material can be used to build homes, home accessories, solar panels, and more. This material allows for more natural lighting sources, which decreases the dependency on artificial lighting sources (Pemberton)	
Stage	Research	
Beneficiaries	Solar Panel Manufacturers, Construction companies	• Cost-effective wood alternative • Biodegradable • Eco-friendly

Case Study 5 ⟪ GIATEC

SmartRock: Monitoring Construction Through IoT

- Countries of implementation: Global
- Company: Giatec Scientific

Company Brief: Giatec Scientific Inc. is a manufacturer of smart testing technologies in the construction industry, based in Ottawa.

Technology	Internet of things/sensors
Description	SmartRock is a wireless device developed by Giatec Scientific that gets embedded in concrete and is resistant to any harsh conditions it may be exposed to. This sensor monitors concrete curing and hardening using the ASTM-approved testing method. Currently, the company produces three generations of this product, with the latest having the property of dual-temperature monitoring capabilities, which enables users to measure temperature values at two locations simultaneously. This allows users to monitor the temperature difference and is linked to a mobile application to help users monitor such data in real time (Hearns)
Stage	In market
Beneficiaries	Engineers, developers, construction companies

	• Real-time data monitoring
	• Early detection of malfunctioning

Case Study 6 ≋ **Fraunhofer** ITALIA

Robotics on Site

- Country of implementation: Italy
- Company: Fraunhofer Italia Innovation Engineering Center (IEC)

Fraunhofer Innovation Engineering Center IEC is the first Fraunhofer Research Institution in Italy, Bolzano, organized by the Fraunhofer Institute for Industrial Engineering IAO, the South Tyrol Employers' Association, and with the support of the Free University of Bolzano. The center supports small and medium enterprises in the region to gain access to applied research (Fraunhofer in Europe).

Technology	Robotics
Description	Engineers at the IEC are in the process of developing and adapting the Husky A200 a four-wheeled, medium-sized robot into an on-site helper for contractors to carry their tools and resources. Ultimately, the goal is to provide automatic navigational properties and smooth navigation on-site and handle the versatile and challenging environmental differences of construction sites (Phillips)
Stage	Testing
Beneficiaries	Construction engineers field workers	• Minimized physical injury associated with load bearing

Case Study 7

Augmented Reality (AR) Architectural Designs

| Sourcing | Material Processing | Design | Construction | Sales & Marketing | Occupation/ O&M | Waste Management |

- Country of implementation: Egypt
- Company: ARki

ARki Design. Architectural interiors offer design and construction management services for residential, commercial, mixed-use buildings, and interior fit-out projects.

Technology	Augmented reality (AR)
Description	ARki is a real-time augmented reality (AR) visualization service for architects. It incorporates AR technology for architectural designs in providing 3D models. This allows architects to design and showcase their designs for their clients through a downloadable application on both iOS and Android devices. The application allows 3D models to overlap onto existing 2D floor plans. Models and designs can be captured and recorded by users for specific views, and scenes, as pictures and video formats to be shared on social media (Lidija)
Stage	In market
Beneficiaries	Real estate developers

Case Study 8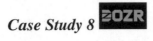

Sharing Economy for Construction

- Country of implementation: North America
- Company: Dozr

DOZR is an online platform that connects construction equipment providers with renters.

Technology	Platforms	
Description	Dozr is a sharing economy-based platform that was founded in 2015 that specializes in the short- and long-term rental of heavy equipment in North America. The platform aims at connecting business owners with excess equipment or equipment not being used at full capacity with renters. The platform is a competitive one-stop shop model.	
Stage	In market	
Beneficiaries	Contractors, developers	• Competitive pricing • Cost management • Optimized of resources

Case Study 9

KONE Care ™ Elevator

| Sourcing | Material Processing | Design | Construction | Sales & Marketing | Occupation/ O&M | Waste Management |

- Countries of implementation: Global
- Company: IBM

IBM is an ICT company headquartered in New York, USA, and is one of the leading technology companies worldwide. The company manufactures and sells computer hardware and software and offers infrastructure services and hosting services and consulting services in different ICT areas.

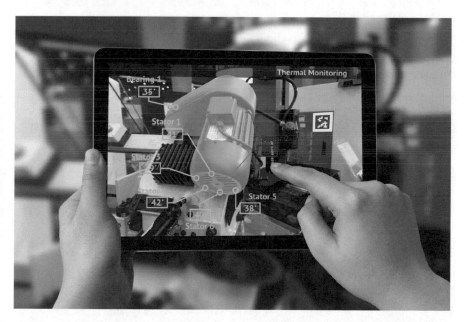

Technology	Artificial intelligence (AI)/predictive analytics (PA), internet of things (IoT), platforms
Description	The Watson IoT platform is software that has multiple applications. KONE, one of the world's top engineering companies, has applied this technology to introduce a new customizable, full-time connected maintenance service for elevators. This enables KONE to predict any potential faults The application gathers its data from the elevators using sensors. This is then constantly monitored, analyzed, and synthesized in real time, allowing operators to view and detect faults and so improve overall performance (Oyj)
Stage	In market

Beneficiaries	Building owners, facility managers, developers	• Real-time prediction of equipment malfunction • Increased safety and reliability in elevators

Case Study 10

Humanoids

Sourcing	Material Processing	Design	Construction	Sales & Marketing	Occupation/ O&M	Waste Management

- Country of implementation: Japan
- Company: Japan's National Institute of Advanced Industrial Science and Technology (AIST)

The AIST is one of the largest public research organizations in Japan, focusing on the practical implementation of technologies in industries and technology commercialization.

Technology	Artificial intelligence (AI)/robotics
Description	The AIST has been developing and advancing humanoids robotics since the early 2000s. The typical robot replaces everyday construction laboring tasks, such as Recognizing different objects and proceeding the correctly carrying them, even when piled up Placing different objects in their rightful place, even though the view is blocked (e.g., due to the object length covering their camera) Recognizing nearby tools that will help them fixate the object in the desired place
Stage	Research & development
Beneficiaries	Construction workers	• Labor efficiency • Safety on-site

Case Study 11

VR Experience in Training

- Country of implementation: USA
- Company: Bechtel

Bechtel Corporation is an American engineering, procurement, construction, and project management company, founded in San Francisco, California, and is one of the largest construction companies in the USA (Engineering, Construction & Project Management).

Technology	Virtual reality	
Description	Bechtel has developed its virtual reality (VR) training program for crane operators to ensure quality, cost-effective training by technical experts. The program is comprised of immersive virtual experiences with game-like controls. Bechtel's industrial expertise has allowed them to craft the content and training module to provide in-depth real-life scenarios and scenario variations	
Stage	In market	
Beneficiaries	Crane operators	• Real-life training • Decreased potential for on-site error • Cost-effective training

Case Study 12 UBITQUITY

Blockchain for Real Restate and Land Records

Sourcing   Material Processing   Design   Construction   Sales & Marketing   Occupation/ O&M   Waste Management

- Country of implementation: Brazil
- Company: Ubitquity, LLC

Ubitquity, LLC is a pioneer in blockchain application in real estate, based in the USA, Delaware. The company's main focus is on the real estate sector stakeholders, governmental bodies, and resellers for smart and safe tracking (About – Ubitquity).

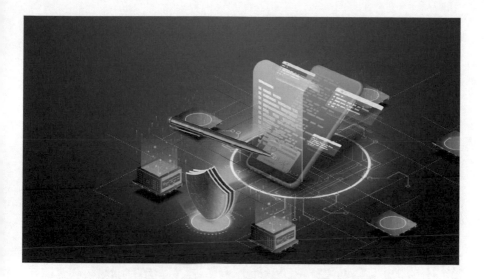

Technology	Blockchain/software-as-a-service (SaaS) platform and application programming interface (API)
Description	Ubitquity employs blockchain technology to strengthen tracking and recording systems of important documents, such as governmental records, contracts, and more. The use of blockchain has added speed and transparency to filing and transactional processes. In 2017, Ubitquity partnered with Brazil's Real Estate Registry Office to improve its land and property ownership recording process. Due to high levels of corruption and fraud, the company's solution was applied to complement the existing system by increasing efficiency and transparency (Ubitquity)
Stage	Pilot (in market)
Beneficiaries	Governments and land owners	• Centralized system for governmental land management • Decreased fraud

Case Study 13 OWTM

MTWO: 5D BIM

- Countries of implementation: Global
- Company: RIB Software

RIB is a software company that specializes in innovative digitization solutions for the construction industry.

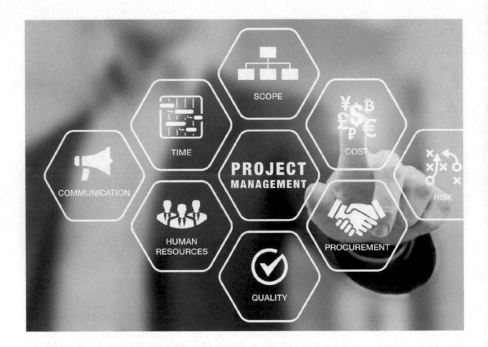

Technology	Platforms/BIM cloud, artificial intelligence (AI)
Description	MTWO is RIB's cloud-based platform that combines five-dimensional building information modeling and artificial intelligence (AI) to provide construction stakeholders, from contractors and project managers to investors, with a holistic platform that allows them to manage projects in real time. The platform can be set up within 48 hours or less, allowing users to set up project plans, monitor real-time progress, and determine the scope of their project, including implementation, prices and costs, quality, and safety. It also allows users to solicit statistics, project-related metrics, scheduling, and quality assurance (MTWO Go Live in 48 hrs)
Stage	In market
Beneficiaries	Project developers Construction companies Industrial companies Contractors and Investors	• Integrated, one-stop platform to manage projects • Increased efficiency • Easier planning and collaboration • Accessibility for all project stakeholders

Case Study 14 sensohive

Maturix

- Countries of implementation: Europe
- Company: Sensohive

Sensohive is a technology-based startup that specializes in wireless sensors and the internet of things

Technology	Internet of things (IoT) and platforms/concrete curing sensor	
Description	Maturix is software that allows the monitoring of concrete using IoT-embedded sensors (Maturix). These sensors are embedded during casting to allow monitoring of the material in real time. This happens through SMS or in-app notifications, connected to a Maturix dashboard (Maturix)	
Stage	In market	
Beneficiaries	Construction companies	• Decreased material waste • Decreased energy consumption in material processing • Reduced heating for concrete curing

Education

Abstract The educational industry comprises establishments such as schools, universities, and training centers that supply knowledge, skills, values, and training in various fields. Education is an integral part of our development because it is our future. Without knowledge, we will be living in a world without advancements and innovation. Education helps society to develop opportunities to attain a better standard of living. A country's prosperity, economy, and social growth directly depend on its populations' literacy rate. Education is a platform that helps people understand the world's problems and the foundation that helps us solve these issues. The education industry's role is to educate people to certify and successfully execute their tasks in the economy and teach the community its principles and morals.

Keywords Interactive board · Learning software · Attendance tracking and monitoring · Interconnected devices · Environment sensors · Data analysis · Learning management system · Data mining · Cloud platform · Open-source learning · Virtual experience · Discovery-based learning · Customized systems · Progress tracking · Inclusion · Process automation

The educational industry comprises establishments such as schools, universities, and training centers that supply knowledge, skills, values, and training in various fields. Education is an integral part of our development because it is our future. Without knowledge, we will be living in a world without advancements and innovation. Education helps society to develop opportunities to attain a better standard of living. A country's prosperity, economy, and social growth directly depend on its populations' literacy rate. Education is a platform that helps people understand the world's problems and the foundation that helps us solve these issues. The education industry's role is to educate people to certify and successfully execute their tasks in the economy and teach the community its principles and morals.

The educational industry, like any other industry, faces challenges which it tries to overcome. One of the most critical being the inconsistency of the quality of education and educational process across various institutions and even within the same institution. Another challenge is the resistance to change and preparing students for the new world. Nowadays, many technological platforms are on the rise, such as smartphones and social media. Institutions need to adapt and help students prepare for the evolving environment by integrating digital literacy or teaching technical skills like coding.

Additionally, an ongoing challenge within the educational sector is catering to disadvantaged and underperforming students, especially in public schools. Many families register their children in public schools due to financial constraints; unfortunately, due to limited resources and funds, public schools do not accommodate students with special needs.

In this digital era, the industry can automate and digitalize educational processes such as registration and tuition payment. It has become easier to attain information, making the educational process more efficient and productive. Additionally, technology in its many forms has helped educational establishments gain a competitive edge by facilitating online classes, assignment completion, and even taking assessment exams all through the internet. Having easy access to materials online provides a massive opportunity for educational institutes to provide global services, thus potentially increasing their revenue. Technological advancements have transformed the educational experiences and opportunities, and its integration in educational institutes has become vital in the students learning journey (Educational Challenges and Opportunities of the Coronavirus (COVID-19) Pandemic | World Bank Blogs)

Value Chain

The value chain for education consists of six phases. The first three are directly associated with educational services and include activities linked to assessment, student selection before admission, the development of teacher skills, and delivery of educational resources. The last three phases are primarily concerned with the community, offering support to the student body, executing marketing materials promoting academic institutions, and maintaining alumni relations.

Case Studies

Case Study 1

IoT for Interactive Learning

| Students Admission | Development of teaching & research skills | Student Learning and Assesment | Student on campus support operations | Marketing | Alumni relations |

- Countries of implementation: Global
- Company: Smart Technologies

Smart Technologies is an international technology company specializing in technology solutions to educational institutions, businesses, and governments. The company was founded in 1987 and introduced the first interactive whiteboard in 1991 (About Smart Technologies | Smart Technologies).

Technology	Internet of things (IoT)/interactive IoT	
Description	Smart Technologies offers an Interactive whiteboard display in the classroom that allows educators to share content while writing on the board or surfing the internet and provide notes online. Students can fully interact on the whiteboard. The SMART board comes with learning software that makes the education process more interactive and customizable based on students' needs (Inspiring classroom experiences| Smart Technologies, 2020). The High School of the Future in Pennsylvania has capitalized on the features of the SMART board to offer personalized education to meet students' needs in the SEN classroom. Educators use visuals to accommodate various abilities as the board can display texts and images for those who cannot hear and voice for those who cannot see (Meeting the needs of all students in a special-needs classroom | Smart Technologies 2019)	
Stage	In market	
Beneficiaries	Educators	• Interactive educational platforms • Ability to save class notes and data • Access to stored data online
	Students	• Engaging and motivational learning • Eliminates note-taking • Lessons stored online
	Parents	• Access educational content

Case Study 2

IoT for Attendance Tracking

Students Admission — Development of teaching & research skills — Student Learning and Assesment — Student on campus support operations — Marketing — Alumni relations

- Countries of implementation: Great Britain and Pakistan
- Company: Innovative Solutions & Development

Innovative Solutions & Development is a relatively small technology solutions company based in Pakistan and the UK. The company was established in 2017, and it was able to offer various solutions with a focus on education and hospitals (Innovative Solutions & Development \| LinkedIn).

Technology	Internet of things (IoT)/cloud IoT
Description	Innovate Solutions and Development offers an attendance tracking IoT device that scans the student's magnetic strip, bar code, or biometric data. This system is called attendance monitoring system. Students have an RFID tag (radio frequency identification – a wireless way to establish identity) that is scanned to the RFID reader of the device. The device captures that data and compares it with the student's stored data in the database, and attendance data is added. Attendance data is stored on the cloud and can be instantaneously shared with students, parents, and educators. Warnings can be automatically sent to students in case the maximum number of classes is missed. Also, detailed reports can be extracted when needed (RFID and IOT for Attendance Monitoring System| Matec Conferences 2017). This can also be applied for employee tracking and bus tracking. One of the systems this company has created is smart attendance system, a digital system used in an educational institute and large organization. The method entails placing an RFID chip in the member's ID. The scanner scans the ID and instantly identified the member. An SMS is then sent to parents, informing them whether their child is present or not. (How it Works| Smart Attendance Solution)
Stage	In market

Beneficiaries	Educators	• The automated attendance tracking system • Accurate data • Generates real-time reports • The decreased human error resulting from redundancy • Increased security and confidentiality
	Students	• Hassle-free attendance tracking • Access to real-time data of students • Receive warnings when students miss classes
	Parents	• Access to real-time data of students • Receive alerts when students miss classes

Case Study 3 INTEGRA SOURCES

IoT for Campus Safety and Security

- Country of implementation: Russia
- Company: Integra Sources

Integra Sources is an IT solutions company that was founded in Russia in 2013. The company's primary focus is on customizable IoT solutions in all industries. They provide IoT solutions based on simple sensors or smartphones to collect and share real-time data online as needed (Company | Integra Sources).

Technology	Internet of things (IoT)/network of IoT	
Description	Integra Solutions offers an integrated IoT security solution that is based on sensors and applications to develop a network of interconnected devices across a school or college campuses so data can be collected, exchanged, and analyzed whenever needed. Installing systems that monitor and track student's behavior, such as connected surveillance cameras, wireless door locks, facial recognition systems, etc. tracks any abnormal behavior and sends alarms when needed. This increases the security of students on campus. Also, adding sensors in the system to monitor the environment, such as temperature and humidity, and sending alerts whenever needed increases students' safety on campus. Monitoring the situation can also increase the efficiency of the schools' resources (IoT Solution Development Services for Education| Integra Sources 2020)	
Stage	New to market	
Beneficiaries	Administration	• Higher security on campus • Reduction in security staff salaries as their function might become obsolete • Smart monitoring of the campus
	Students	• Safer and more secure campus
	Parents	• Safer and more secure campus • Access to real-time data

Case Study 4

Visual Analytics for Capturing Gaps

| Students Admission | Development of teaching & research skills | Student Learning and Assesment | Student on campus support operations | Marketing | Alumni relations |

- Country of implementation: USA
- Company: Tableau

Tableau is a data visualization company that was launched in 2003. The company aims to provide fast and secure data analysis by visualizing the data. It is a company that produces intelligent analytic software that can convert data to actionable insights. The company offers free products available for everyone, such as Tableau public and other paid products with advanced features and higher security, such as Tableau desktop (About: Tableau Mission I Tableau)

Technology	Big data and advanced analytics/visual analytics	
Description	Tableau visualization software works on analyzing data from various sources such as databases, cloud databases, spreadsheets, etc. The software generates easy-to-understand visualizations that quickly assist the user in spotting trends, opportunities, and gaps. For prominent universities, such as Indiana University, which has more than 100 thousand students from all over the world, analyzing their data through Tableau helps the university gain a more insightful picture of their admission and scholarship strategy. This data may also be used to determine the target audience for marketing purposes. Other universities, such as Santa Barbara Community College, use data from academic catalogs and enrolment to forecast the section offerings, predict gaps within course offerings, and update the course descriptions as needed. Theoretical data analysis of enrolled students assists universities in identifying at-risk students and proactively reach out to them. One of Tableau's unique features is that it provides an overall view of the data and breaks it down as needed in seconds (Indiana University and Santa Barbara City College help students succeed with course-level ⏐Tableau Solutions 2020)	
Stage	Available on market	
Beneficiaries	Administration	• Effective decision-making • Effective resource utilization • Effective forecasting
	Students	• Easier registration process • Updated course offerings
	Educators	• Updated course content to meet market needs

Case Study 5 CLOUDΞRA

Predictive Analytics for Better Decision-Making

- Country of implementation: USA
- Company: Cloudera

Cloudera is a company that offers software for data warehousing and data analytics. Its main objective is to turn a large amount of data into insights that help organizations to solve problems it faces. The company was founded by former employees in the top IT leading industries such as Google, Oracle, Yahoo, and Facebook (About ⏐ Cloudera).

Technology	Big data and advanced analytics/predictive analytics and machine learning
Description	Cloudera offers multifunction data analysis tools that read from multiple data sources and identify patterns, forecasts, gaps, etc. Florida University uses Cloudera to analyze the academic data stored on the University's learning management system. Cloudera uses specific algorithms such as "Apache Spark Machine Learning" to predict if a student will pass the course or not based on historical data. The university proactively gets in touch with those expected to fail. Florida University's testing center also uses Cloudera to predict demand on seasonal tests offered by the university. Previously, the university had no idea how many students will show up for a specific exam, which led to inefficiency. With Cloudera, the university can predict seasonal test demand to better schedule their tests and provide the needed resources accordingly (Florida State University: Boosting student success and powering decision-making through big data | Cloudera 2020)
Stage	New to market

Beneficiaries	Administration	• Improved student success rate • Higher administrative efficiency • Efficient decision-making
	Students	• Proactively supports students at risk of failing

Case Study 6 SPSS

Data Mining for Higher Efficiency

| Students Admission | Development of teaching & research skills | Student Learning and Assesment | Student on campus support operations | Marketing | Alumni relations |

- Country of implementation: USA
- Company: SPSS

SPSS is a software company specialized in predictive analytics that aims to find conclusions about the current status and future events. It was founded in 1968 and was acquired by IBM in 2009 to become IBM SPSS (SPSS Inc. | Wikipedia)

Technology	Big data and advanced analytics/data mining
Description	SPSS is a powerful tool that analyzes a vast amount of data using sophisticated statistical procedures. This could be useful for the university's alumni data, which could be at least ten times larger than the enrolment figures. Communicating with this massive number of alumni is very costly and most likely not worth it since engagement from these alumni is relatively low. Relying on a tool like SPSS will help the university identify alumni with high engagement probability through data mining. A more useful list of alumni is determined through data mining instead of communicating with all alumni, which is more expensive and inefficient (Data Mining Applications in Higher Education ISPSS 2020)
Stage	New to market

Beneficiaries	Administration	• More effective communication • Cheaper communication
	Alumni	• Avoid getting irrelevant messages

Case Study 7 **Linked in** Learning

Platforms

- Country of implementation: USA
- Company: LinkedIn Learning (Lynda previously)

LinkedIn Learning is a platform that offers various online video courses that are provided by industry experts. These videos train the users on the computer, business, and creative skills. The company was founded in 1995 by Lynda Weinman and was known as Lynda.com. In 2015, LinkedIn, the largest professional network with more than 675 million users, acquired it. In December 2016, Microsoft acquired LinkedIn, so now it is a Microsoft subsidiary (LinkedIn Learning I Wikipedia).

Technology	Content platform/big data and advanced analytics
Description	LinkedIn Learning offers various video tutorials for beginner and advanced levels. Compared to onsite training, which could cost a lot, online instructions are more cost-effective. This is what drove a big university like Georgetown University to start using LinkedIn Learning. The university wanted to train its employees on new technology trends and applications. Off-site training would cost more than $1000, so they decided to replace the off-site training with LinkedIn Learning online training. The staff started using LinkedIn Learning tutorials, and students began using LinkedIn Learning to get more assistance in-class projects. They have also used LinkedIn Learning for career development. (Georgetown University provides the entire campus with cost-effective lynda.com instruction I LinkedIn learning)
Stage	Available on market

Beneficiaries	Administration	• Cost-effective training solutions • Flexible training packages
	Students and alumni	• Career Development support • Training support

Case Study 8

Communication Platforms

| Students Admission | Development of teaching & research skills | Student Learning and Assesment | Student on campus support operations | Marketing | Alumni relations |

- Countries of implementation: Global
- Company: Zoom

Zoom is a US leading company in the video conferencing industry. The company was founded in 2011, aiming to collaborate between teammates. Zoom is considered the "original software-based conference room solution" (About| Zoom 2020).

Technology	Platforms/communication
Description	Zoom offers a reliable cloud platform for audio and video calls; it also allows for collaboration and chatting between users and hosting webinars. Zoom has been used in many well-known schools and universities in Egypt and abroad, such as UMASS, Yale, AUC, and AIS
Stage	In market

Beneficiaries	Educators	• Remote teaching • Optional class recording • Virtual collaboration
	Students	• Access to recorded and live lessons • Virtual participation and collaboration

Case Study 9

Learning Platforms for Global Reach

Students Admission ➤ Development of teaching & research skills ➤ Student Learning and Assesment ➤ Student on campus support operations ➤ Marketing ➤ Alumni relations

- Countries of implementation: Global
- Company: Moodle

Moodle is an online free, open-source educational platform that easily allows professors to create and customize their workspace. The company aims to provide professionals with a platform that will enable content creation on remote classes that are interactive and engaging. The company was founded in 2001 with a mission to "empower educators to improve the world" (Aboutl Moodle 2020).

Technology	Platforms/learning management systems	
Description	Moodle is an open-source learning platform meaning that users can download the source code, then customize the platform, and enhance as needed. Universities, such as Zagreb in Croatia, have capitalized on Moodle to serve the university and higher education community in Croatia. The Computer Centre at the University of Croatia (SRCE) is responsible for managing the application of ICT in learning for higher education in Croatia. SRCE was aiming to develop a process of E-learning Implementation. This included providing a virtual learning platform for various users across the country. SRCE customized Moodle's platform and availed it to all users through a single sign-on to eliminate the challenge previously stated. With a single installation, the platform was used by more than 100 institutions in Croatia, where educators can easily share their content (University of Zagreb University Computing Centre SRCE| Moodle, 2019)	
Stage	In market	
Beneficiaries	Administration	• Cheap and easily accessed platform • Higher adoption for E-learning • Unified E-learning process
	Professors	• Accessible E-learning platform • Ease of content sharing
	Students	• Accessible educational content • Option to translate to other languages

Case Study 10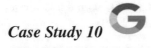

Virtual Reality for Unique Admission Process

- Country of implementation: USA
- Company: Google

Google is an American market leader that offers various internet-related products and services such as search engines, cloud computing online advertising, as well as different software and hardware IT-related products. The company was founded in 1998 and is currently considered one of the biggest IT companies (Google| Wikipedia 2020).

Technology	Virtual reality/virtual reality headset (Google Cardboard)	
Description	Google Cardboard is a virtual reality tool developed by Google that allows users to experience virtual reality on a smartphone through foldable cardboard. It is an affordable and accessible way to experience virtual reality. Google Cardboard has been used by the Enrollment Management Office in NYU's Tandon School of Engineering to provide newly admitted students with a glimpse of campus life before students enter the campus. Tandon School of Engineering has created a virtual reality mobile application that shows a quick overview of the campus and an introduction to student life. Newly accepted students received a package in the mail with a Tandon branded cardboard and instructions on how to use the cardboard and download the mobile application. The VR application developed was later used for different purposes such as taking students on a virtual trip to Mars (NYU's Tandon School of Engineering enrolls more women with the help of VR apps I Google 2020)	
Stage	In market	
Beneficiaries	Administration	• Automated campus tours that do not require staff efforts • A higher number of applicants if applicants can experience campus life virtually in advance
	Students	• Access to campus tours whenever needed • Access to other virtual experiences that could have never experienced if virtual reality did not exist

Case Study 11

Augmented Reality for Interactive Education

| Faculty Recruitmenet | Students Admission | Development of teaching & research skills | Student Learning and Assesment | Student on campus support operations | Marketing | Alumni relations |

- Countries of implementation: UK, USA, and Singapore
- Company: Blippar

Blippar is a London-based, augmented reality company that was founded in 2011. The company's primary goal is to develop augmented reality systems that are engaging and entertaining. The company currently has various branches across the world. Blippar was the first company to introduce an AR digital advertising platform (Blippar| Wikipedia, 2020).

Technology	Augmented reality/marker-based augmented reality
Description	Blippar app is a powerful augmented reality (AR) application that allows users to experience the world around them through a smartphone. The Blippar application can help students learn about subject matters through "discovery-based learning," where they can view detailed educational content. For example, if they choose a volcano, they can see it erupt.
	Blippar additionally allows students to experience how objects move in real life using 3D technology. The Blippbuilder, a JavaScript developed by Blippar that allows developers to build augmented related experiences, can create interactive books. Gamification of learning can also be done through an augmented reality where students learn by games, such as treasure hunting (Bringing the Augmented Reality Education to Your Home for Free! Blippar 2020)
Stage	In market
Beneficiaries	Students

- Interactive learning
- Engaging content
- Better retention of concepts
- Access to unique learning experiences

Case Study 12 ▼CTI

Artificial Intelligence in Education

- Countries of implementation: Global
- Company: Content Technologies

Content Technologies is a company that specializes in the growth and research of artificial intelligence platforms. The company aims to provide businesses with tools that will help them find intellectual ideas and content (Content Technologies, Inc.).

Technology	Artificial intelligence(AI)	
Description	Content Technologies is a company specializing in the digitalization of business processes with the use of artificial intelligence (AI) platforms. This company has created an artificial intelligence (AI) technology, Cram101, to break the contents of a book into a comprehensible "smart study guide." The technology may include summaries on essential sections in the book and practice problems. In other words, Cram101's online artificial intelligence (AI) services read the content of books, summarizes them, and emphasizes critical concepts within the chapters. The platform can do all of that in 1 hour (Examples of Artificial Intelligence (AI) in Education I Emerj 2019)	
Stage	In market	
Beneficiaries	Students	• Interactive and productive learning • Time-saving

Case Study 13 **CARNEGIE LEARNING**

Smart Tutoring System

- Country of implementation: USA
- Company: Carnegie Learning

Carnegie Learning is a company in the education industry specializing in transformational math and aims to provide educators, scholars, and professionals a more enhanced mathematic literacy. The company has aided students and professionals with its cutting-edge research software, books, and analysis services (Carnegie Learning Partners with OpenStax to Offer the Most Powerful Affordable College Math Solution on the Market | Business Wire 2017).

Technology	Artificial intelligence(AI)/cognitive technologies	
Description	Carnegie Learning has created an artificial intelligence (AI) software platform called Mika to give students, especially those in higher education, a more customized tutoring system. It aims at college students who may need remedial learning to support their degree achievement. The company proved that remedial institutes receive billions yearly but have a low success rate in math-related courses. With this app, students, especially college students, have access to help at a low cost (Examples of Artificial Intelligence (AI) in Education	Emerj 2019)	
Stage	In market	
Beneficiaries	Students	• Cost-effective alternative to remedial courses • Personalized learning platform

Case Study 14

Artificial Intelligence (AI) Revolutionized the Education Sector

| Students Admission | Development of teaching & research skills | Student Learning and Assesment | Student on campus support operations | Marketing | Alumni relations |

- Countries of implementation: Global
- Company: Thinkster Math

Thinkster Math is a company that focuses on customized, technology-based math tutoring using data analytics. Their patent Active Replay Technology (ART) application allows tutors to review students' works and give remarks on their mistakes. The company aims to develop long-lasting educators with their use of analytical and reasoning skills (Thinkster Math | LinkedIn).

Technology	Artificial intelligence(AI)/machine learning
Description	Thinkster Math created an app that tracks students while they are answering math problems using artificial intelligence (AI) and deep learning technologies. Through this app, students solve mathematical problems, and the software produced a detailed analysis of the students' progress and the skills that were used during the test. Educators and professors used this tool to identify how students understand different math topics and understand students' weak spots without going through a massive number of assignments and tests, and this gives educators more time to develop customized assignments to help students grow and improve their weaknesses (4 Ways Artificial Intelligence Revolutionizing Education| Dell Technologies 2018)

Stage	In market	
Beneficiaries	Students	• Improved academic success • Identify learning challenges • Design assignments to support learning
	Educators	• Identify and track student achievements • Efficient use of time

Case Study 15

Crowdsourced Learning

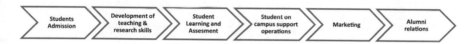

- Countries of implementation: Global
- Company: Brainly

Brainly is one of the world's biggest peer-to-peer learning platforms created to help the community in the education sector. Brainly has websites worldwide where students could answer or ask questions in any field. The platform has over 200 million users per month. The company aims to inspire and promote collaborative learning in a comfortable atmosphere (Brainly, Wikipedia).

Technology	Artificial intelligence(AI)/machine learning	
Description	Brainly is an online platform that helps students learn from each other. Through this platform, peers can ask questions and have them answered by other students. The company uses artificial intelligence (AI) and machine learning technologies to help filter through the questions and answers to ensure the presented product is of high quality, such as eliminating incorrect answers and spam (4 Ways Artificial Intelligence Revolutionizing Education| Dell Technologies 2018)	
Stage	In market	
Beneficiaries	Students	• Easy access to academic support • A more engaging way to share educational information • Access to international support academically

Case Study 16 gradescope

Efficiency in Grading

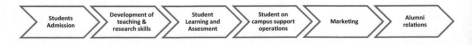

- Country of implementation: USA
- Company: Gradescope

Gradescope is an American-based company that was developed in 2012. A PhD student, Arjun, is the cofounder and creator of this company. Its purpose is to create an automated remote platform to help decrease the time educators spend in correcting exams and assignments while maintaining a high-quality output (Arjun Singh l).

Technology	Artificial intelligence(AI)/machine learning	
Description	Gradescope created a platform that speeds the process of correcting exams and schoolwork for instructors. The student uploads their document on the platform and, with artificial intelligence (AI) technologies, the platform can categorize and assemble answers to check several works at once. The company states that the time to grade and correct students' work has been reduced to 70%. The platform is adapted with software that provides educators with a thorough analysis of the students' work, highlighting their strengths and weaknesses (ARTIFICIAL INTELLIGENCE(AI)in Education Use Case #7: Gradescope l Disruptor Daily)	
Stage	In market	
Beneficiaries	Educators	• Reduced marking time • Identifies student challenge areas • Provides customized assignments to support learning
	Students	• Personalized assignments tailored to support individual learning needs

Case Study 17 ↰DAQRI

AR to Enhance the Education Experience

| Students Admission | Development of teaching & research skills | Student Learning and Assesment | Student on campus support operations | Marketing | Alumni relations |

- Countries of implementation: Global
- Company: DAQRI Studio

DAQRI Studio is an AR-based, small-scale company focusing on AR applications and how we can integrate such technologies in our surroundings. Its main objective is to provide a more productive and effective solution within the workplace (About DAQRI I DAQRI).

Technology	Augmented reality
Description	DAQRI studios have created an application called Elements 4D that is downloadable on both IOS and Android devices. The application's purpose is to help students in any course material related to chemistry in a virtual, remote manner. The app provides different chemical compounds that can be mixed and simulate the results to match the final output in real life. This is a good option for students to understand chemical concepts and reactions remotely during the COVID-19 pandemic (Augmented Reality in Education I ThinkMobile).
Stage	In market

Beneficiaries	Students	• Virtual, interactive, and engaging learning
	Educators	• High-quality distance education

Case Study 18 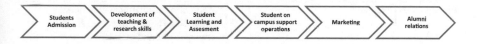 Microsoft

Classroom Applications

Students Admission	Development of teaching & research skills	Student Learning and Assesment	Student on campus support operations	Marketing	Alumni relations

- Countries of implementation: Global
- Company: Microsoft

Microsoft is an American corporation that specializes in computer technology. It is the largest and most significant software company worldwide. It develops, builds, and sells software, electronics, computers, and any computer-related services. The company aims to create top-notch platforms and software that help empower people and different organizations (A Brief History of Microsoft - The World Biggest Software Company I DSP).

Technology	Artificial intelligence/cloud technology	
Description	Seeing artificial intelligence (AI) application was developed by Microsoft and can be downloaded on any IOS device. The free app aims to help the visually impaired. With the use of the device's camera, it identifies what the camera sees and states it for the user. The app can read out texts, describe products, and report currency details. It can describe a person by saying their age, height, gender, and mood. This is useful for visually impaired students to have interactions in the classroom or playground and then "read aloud" to them (Artificial Intelligence in the Classroom| Microsoft Education Blog 2018)	
Stage	In market	
Beneficiaries	Students	• Liberating educational experience • Increases visually impaired student engagement and interaction • Enhances academic outcome

Case Study 19

Blockchains Integration in Education

- Country of implementation: USA
- Company: Hyland

Hyland is a software company that developed a business content management and process management software suite called OnBase. Those software applications are used in many industries like health care, banking, and education. It specializes in content services that help in the efficiency and productivity in the workplace (Hyland Software | Wikipedia).

Technology	Blockchain
Description	Hyland created certificates, called Blockcerts, that could be issued digitally, attested, and verified by the issuing establishment. This has created a more efficient way of releasing and verifying certificates using blockchain technologies. It helps diminish any dishonesty relating to academic credentials. Blockcerts are signed digitally by the organization but still belongs to the user. The difference between Blockcerts, blockchain-based certificates, and paper certificates is that documents can be verified by comparing the Blockcerts to the "digital footprint" found in the blockchain (Blockcerts in Best: FAQ | Best 2019)
Stage	In market
Beneficiaries	Students

Beneficiaries	Students	• Eliminates extending and tedious credential authentication
	Professionals	• Fast and reliable credential check

Case Study 20 **LANIER**

Automation in Admissions

Students Admission	Development of teaching & research skills	Student Learning and Assesment	Student on campusupport operations	Marketing	Alumni relations

- Country of implementation: USA
- Company: Lanier

Lanier is a subsidiary of Ricoh Corporation, a Japanese multinational electronics company, and is one of the largest providers of management solutions worldwide. It aims to provide clients with solutions to enhance efficiency and minimize business costs (Lanier Worldwide, Inc. | Reference for Business).

Technology	Automation and digitalization	
Description	There are millions of annual college applicants. Colleges face a challenge processing all the presented documents and certificates. Lanier has provided colleges with an admissions process automation solution. This technology helps eliminate administrators' process of managing hardcopy documents and replaces it with automated workflow and data analysis. This will help enhance the processing of student information. Usually, administrators receive reports from students from various sources such as email, fax, and hard copy. It became challenging, labor-intensive, and tedious to process and organize such documents. This service provides the team with a digitalized alternative to paper-based document processing that makes the entire process a lot faster and more effective (Admissions Process Automation | Lanier)	
Stage	In market	
Beneficiaries	Students	• Faster communication with administrators • Confidential
	Administrators	• Diminishes human error • Reduces physical workload • Hastens college admission process

Food and Beverage

Abstract The food and beverage industry is composed of companies, institutions, and establishments that engage in farming, manufacturing, distribution, and the selling of food and beverages. The manufacturing phase includes the processing of raw materials, like farming outputs, meats, and dairy and the production of soft drinks, alcoholic beverages, packaged foods, and other modified foods. The companies in the food and beverage industry process raw materials into food products and then wrap, distribute, and sell them through the different distribution channels, finally reaching the end consumer. Then, the distribution phase involves transporting the finished food product into the hands of consumers which can be done through different channels like retailers and restaurants (ScienceDirect).

Keywords Supply management · Automated alerts · Cashless · Connected · Herd monitoring · Collaborative robots · Lightweight material · Ergonomic design · Satellite navigation · Contactless applications · Delivery processes · Voice recognition · Traceability systems · AI algorithm · Sustainable resources · Information systems · Smart packaging · Customer engagement · 3D experience · Automation · Optical character recognition · Automatic applicator

The food and beverage industry is composed of companies, institutions, and establishments that engage in farming, manufacturing, distribution, and the selling of food and beverages. The manufacturing phase includes the processing of raw materials, like farming outputs, meats, and dairy and the production of soft drinks, alcoholic beverages, packaged foods, and other modified foods. The companies in the food and beverage industry process raw materials into food products and then wrap, distribute, and sell them through the different distribution channels, finally reaching the end consumer. Then, the distribution phase involves transporting the finished food product into the hands of consumers which can be done through different channels like retailers and restaurants (ScienceDirect).

The food and beverage industry is one of the most critical sectors driving the global economy and employing a significant share of the employment force worldwide. Over the past decade, it has undergone changes that could be attributed to the increased level of automation and innovation, as well as the ever-changing

© Springer Nature Switzerland AG 2023
M. Anis et al., *Mapping Innovation*,
https://doi.org/10.1007/978-3-030-93627-3_6

consumer demands and preferences. Hence, the technological advancements and changes in food science are presenting a significant number of opportunities for many food and beverage industries today (7 Global Tech food Trends to watch for in 2019 | Rocket Space, 2019).

Many small companies in the food and beverage industry have faced challenges due to the number of people rejecting processed foods and leaning toward healthier choices with cleaner labels, which shows that a lot of companies are failing to adapt to consumers' preferences and different lifestyles. Moreover, product traceability and company transparency are essential nowadays for many people who might question many companies' choices when it comes to packaging and labels (Challenges and Opportunities in the Food and Beverage Industry | Thomas, 2020); additionally, consumers are now targeting environmentally friendly products to help reduce the consumption of the biodegradable products. More and more companies are striving to make the food and beverage industry environmentally friendly by banning plastic waste and numerous recycling practices (Top 8 Challenges of Food and Beverage Industry to Watch Out for | Global Market Insights 2020).

As for the time of the Covid-19 pandemic, the food and beverage industry was not ready or prepared for such a spike in demand. The lockdowns that took place in almost all countries have prevented employees from working in factories, which in return has impacted the supply chain. The farming and agriculture activities have made a significant hit, and consumers have started to panic buy, which has deepened even further the food shortage (Top 8 Challenges of Food and Beverage Industry to Watch Out for | Global Market Insights 2020). On the other hand, the rise of e-commerce is considered a significant opportunity for the food and beverage industries during Covid-19. With consumers not wanting to leave their houses during this pandemic and risk getting infected, there is a massive rise in sales due to e-commerce resulting from its accuracy and immediacy (The Role of e-commerce in the Food and Beverage Industry | MRI 2018).

Value Chain

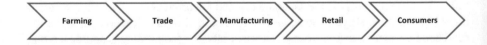

The value chain of the food and beverages industry is characterized by the following activities: farming, trade, manufacturing, retail, and consumption. The first step of the value chain is concerned with growing, harvesting, and preparing the inputs used in the industry. Different types of inputs include fresh produce, dairy, nuts and grains, meats, and poultry, among others. Afterward, players are involved in trade activities that allow manufacturers of the industry to access their required inputs. This stage of the value chain includes manufacturing needs identification, procurement, payment, and delivery of the inputs. Afterward, manufacturers identify customers' needs, carry out R&D activities to identify best practices, and finally add value to the inputs by processing and packaging them according to customers' needs. In the retail stage, the products are distributed for sale to the end consumers via different channels such as wholesalers, retail stores, or restaurants. Finally, consumers review and select products from the available selection of products, purchase their needs, and finally consume the products.

Case Studies

Case Study 1 elnfochips

Intelligent Vending Machines

- Country of implementation: USA
- Company: eInfochips

eInfochips is the leading company provider that designs product engineering solutions and services. It was founded by Paul Shroff in 1994; this company has designed over 500 products with more than ten million units used all over the world. eInfochips has provided and designed solutions for companies in different areas, including health care, industrial automation, aerospace, media and broadcast, and many more.

Technology	Internet of things (IoT/sensors, data analytics, cashless payments, and GPS tracking)	
Description	eInfochips is one of many companies working on the development of an intelligent vending machine. eInfochips has been working on this concept for one of US Fortune 100 companies. This intelligent vending machine uses IoT technology and data analytics to interpret sensor data which will, in turn, send automated alerts to management when the supplies are low. Moreover, the sensors embedded in this system are also used to issue alerts of possible malfunctions, and also if any of the equipment is moved, the company can monitor its location using GPS tracking. Businesses use this new technology to automatically order new products. This innovative vending machine takes into account the ongoing demands of different products and brands and includes features like cashless transactions, smartphone payment, predictive maintenance, cloud services and analytics, customer loyalty programs, and many more. Additionally, this technology can monitor customer behavior patterns and supply coupons and loyalty programs to encourage customers (Connected Vending Quenches Consumers' Thirst 2019)	
Stage	In market	
Beneficiaries	Stakeholders	• Optimize the product mix in any given location while preventing stock-outs • Recognize potential malfunctions and flag possible theft • Lessens downtime • Enables preventive maintenance • Lowers costs • Provide beneficial data on customer use
	Consumers	• Cashless payment • Loyalty program rewards

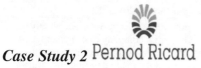

Case Study 2 Pernod Ricard

Connected Bottles

Farming | Trade | Manufacturing | Retail | Consumers

- Countries of implementation: France, China, and the USA
- Company: Pernod Ricard

Pernod Ricard is a company that produces alcoholic drinks and was founded in France in the year 1975. It has produced numerous numbers of prestigious alcoholic drinks, such as Ricard pastis, Martell cognac, and many more. It is the world's second largest wine seller with net sales of around nine billion euros (Joseph 2017).

Technology	Internet of things (IoT)/QR codes and augmented reality (AR)	
Description	Intending to stay connected to its consumers, Pernod Ricard decided to rely on the internet of things (IoT) technology. The spirits company took advantage of its young consumers' always-connected lifestyle to get into and influence the experience of the consumers. They created what they called "connected bottles" which are significantly different from normal bottles in several ways. These bottles have QR codes which, when scanned, give suggestions on where to go with friends, nearest pubs and bars, cocktail recommendations, and what could be eaten or drunk alongside for a better drinking experience. The bottles also have features that allow consumers to check the veracity of a bottle in a store where fake goods are widespread. Moreover, they can also trace any bottle's journey, from the distillery to the shelf, as well as get general information about the brand itself. These bottles collect data on consumer consumption, for example, how, when, and where the consumers are consuming their products (Joseph 2017)	
Stage	In market	
Beneficiaries	Stakeholders	• Increased brand perception • Increased brand loyalty • Stronger business partnerships • Decreased incidence of fake goods
	Consumers	• Enriched experience • Recommendations and brand information • Track and traceability • Increased customer satisfaction

Case Study 3

Smart Agriculture and Dairy Farming

- Country of implementation: Canada
- Company: Joe Loewith and Sons Ltd. Farm

Joe Loewith and Sons Ltd. The farm is a dairy farming industry that is owned and run by Carl Loewith, Dave Loewith, and Ben Loewith. It provides innovative solutions to measure and monitor cows' performances (Vatulan, Quigley, and Masouras 2017).

Technology	Internet of things (IoT)/sensors and GPS tracking
Description	Sensors and certain developed agricultural software allow the monitoring of every cow in the herd. A bracelet or a pedometer is attached which allows for individual identification and provides data about the cows' activity by monitoring and recording the number of steps per hour, milk produced, as well as conductivity levels of the produced milk which could be a sign for the presence of infections in the milk or diseases in the cow. The sensors and the software generate a report that monitors the overall health and needs of each cow. The software is also used for staff accountability, e.g., employees who are responsible for the milking process and the cleaning of the udders can later be held accountable if the technology detects disease in the udders of the cows. Additionally, data collected on the health risks of the cows that could later affect their productivity can be used to support workers' claims (Vatulan et al. 2017)
Stage	In market
Beneficiaries	Farmers

Beneficiaries	Farmers	• Monitoring employee effectiveness • Monitoring herd quality • Increasing productivity • Increased efficiency in the calving process • Cost-effectiveness with precision agriculture • Optimizing the number of chemicals and fertilizers used
	Consumers	• Disease-free milk

Case Study 4

Cobots

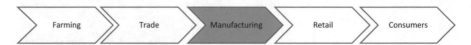

- Country of implementation: Spain
- Company: Atria Innovation

Atria Innovation is a company that designs and develops innovative and technological solutions to help other businesses solve real problems, increase their competitiveness, quality of the product, and profitability of the processes (Atria Innovation).

Technology	Artificial intelligence (AI)/robotics technology		
Description	Collaborative robots (cobots) have evolved significantly since they were first invented as they have now a much more prominent presence in the F&B industry. Cobots, or collaborative robots, are robots intended to interact with humans in a shared space or to work safely in proximity. Cobots stand in contrast to traditional industrial robots which are designed to work autonomously with safety assured by isolation from human contact. Cobot safety relies on lightweight construction materials, rounded edges, and limits on speed or force. Safety may also require sensors and software to assure good collaborative behavior. These cobots have replaced the traditional, huge industrial robots and bulky packaging machines that usually occupy vast spaces in the production facility. Cobots in the F&B industry are designed to work side by side with the employees with no barriers or physical restrictions which makes automation in the F&B industry more convenient, affordable, accessible, and much easier. In addition to the features of user-friendliness, safety, and ergonomic design, these robots can carry a wide range of tasks at much lower costs (Case-Study: Cobots In F&B Manufacturing 2017)		
Stage	In market		
Beneficiaries	Food and beverages manufacturers	• Minimizing production downtime • Time-efficient • Multitasking ability • Cost-efficient as reduced production waste and raw materials • Easy installation and use • Raised productivity • Enhanced overall performance	
	Labor force	• Increased workplace safety • Increased worker engagement and retention	

Case Study 5

Robotics Automation in Agriculture: BoniRob

- Countries of implementation: Germany
- Company: Bosch

Bosch is a company that designs products and technologies and provides services to other businesses to improve quality of life and help protect natural resources. It designs technologies for different areas, such as mobility, home, industry and trades, and many others (Bosch).

Technology	Artificial intelligence (AI)/robotics technology and machine learning	
Description	BoniRob is a robot the size of a small car and is designed to perform different tasks in the farming industry in outdoor fields, such as weed removal and manage plant breeding. By adopting machine learning algorithms and using pictures of the target crop and invasive weeds, BoniRob could be trained to identify and separate the target crop as distinct from invasive plants. BoniRob then used a rod device to crush the weeds thus eliminating the need for herbicides. With accuracy up to 1 cm, BoniRob uses video imaging and satellite navigation to traverse the fields. With the potential of adding communication devices to the robots, these robots could start acting like a fleet interacting with each other for faster more cooperative operation. The robot could also act as a field laboratory collecting valuable information about the plants and the surrounding environment which can be used for research and development purposes (Pittman 2015)	
Stage	Testing	
Beneficiaries	Farmers	• Greater productivity • Higher precision performance • Contactless but a precise chemical application • Faster and more accurate crop harvesting • Mechanical weeding
	Agricultural and farming research and development institutes	• Supports research and development • Provides useful data (soil conditions, the effect of fertilizers, climate changes, etc.)

Case Study 6

Robotic Fulfillment System: Alphabet

| Farming | Trade | Manufacturing | Retail | Consumers |

- Country of implementation: USA
- Company: Walmart Inc.

Walmart Inc. is an American multinational operator and over the last 50 years has become one of the largest retailers in the world. Nearly 265 million people visit around 11,500 stores and e-Commerce websites each week in 27 countries (Walmart Inc.).

Technology	Artificial intelligence (AI)/robotics technology
Description	Walmart has developed its automated storage and retrieval system "Alphabot," a step in the path to revolutionizing the grocery pickup and delivery processes. Alphabot is used to get refrigerated and frozen products ordered online using a self-driven car then assembles them and delivers them to a workstation. An employee then checks, bags, and delivers the final order which makes the retrieval process much faster and easier. Alphabot is a mobile robot that can operate in a multilevel storage structure. A number of these robots can be used together in a distribution center forming a fleet of robots which are all controlled by a single control unit that receives the orders electronically and signals the fleet of robots to do their tasks according to the map of the center and the distribution of the items in the aisles and racks. Alphabot robot systems eliminate the need for using any material handling tools, lifts, or conveyors. It also eliminates the need for human interference in the process of item retrieval and handling (Stallbaumer 2020)
Stage	In market
Beneficiaries	Retailers	• Less dispensation time • Increased accuracy of product retrieval from inventory • Time-efficient, especially during peak times • Smother and more reliable warehouse operations • Faster order preparation • More capital efficiency in operations
	Consumers	• Faster and more accurate delivery of ordered items

Case Study 7

Amazon Fresh

- Country of implementation: USA
- Company: Amazon

Amazon is an American multinational technology company and is one of the most valuable companies in the world. Its services include e-commerce, cloud computing, digital streaming, and artificial intelligence (AI). It is one of the biggest players in the tech field worldwide and one of the most powerful companies culturally as well as economically. Over the years Amazon succeeded not only in penetrating the tech field but also in totally disrupting and changing the way this field was tackled using innovation. Currently, Amazon is the world's largest online marketplace, internet-based company, and AI-assistant provider both in terms of revenue and market value (Amazon).

Technology	Artificial intelligence (AI)/voice commerce
Description	Over the past years, Amazon continued to expand Amazon Fresh, its grocery delivery service. With the recent expansions, Amazon developed their voice recognition smart home speakers and announced that grocery shopping can be made with the help of amazon's famous artificial intelligence (AI) assistant Alexa. Using learning algorithms and AI technology, Alexa can provide better assistance to shoppers based on their previous shopping experiences and shopping carts, as it remembers their favorite items and recurring product selections for an easier and faster shopping experience (Johnston 2019)
Stage	In market
Beneficiaries	Retail stores

	Retail stores	• Increased sales • Customer data identification
	Users	• The convenience of using vocal commands to make a purchase • Faster ordering • Personalized buying experience

Case Study 8

Blockchain-Based Food Supply Chain

- Countries of implementation: USA
- Company: IBM

IBM (International Business Machines) was founded by Charles Ranlett Flint in the year 1911; it is a global company that designs technologies using computing, software, cloud-based, and hardware services to solve the world's challenging issues (Computer Business Review).

Technology	Blockchain/QR coding
Description	As consumers buy their food items from retailers, it is not very rare that a food product turns out to be spoiled or unmatching to the typical hygiene or health standards. This occasion usually requires intermediate intervention from the retailer and the manufacturer as well to pull out the product from the market and to open an investigation to trace back to the origin of the problem across its value chain. In these cases, a full record of the route of the food product across the value chain is needed starting from the farm or the raw materials till the retailer selling the product to the consumer (Donaven 2019). IBM launched its Food Trust platform which is a blockchain technology-based traceability system across the whole F&B supply chain that could be used by retailers, manufacturers, and suppliers in the F&B industry. Walmart in turn was among the first companies to partner with IBM on the project and announced that it requires all its suppliers to implement this traceability option on their products as a condition for Walmart to accept them. Using this blockchain technology and through a simple QR code scan on any food product, this product could be traced back to its origin in a matter of seconds; we could identify where it was grown, manufactured, stored, and how long each of these processes lasted (IBM)
Stage	In market

Beneficiaries	Consumers	• Increased food safety • Increased satisfaction • Accuracy in food labeling
	Stakeholders	• Less liability exposure on retailers • Better crisis handling in case of food recalls • Increased trust among the different blocks of the value chain • Reduced cost of problem allocation (the exact batch with the problem could be identified)

Case Study 9

PuduBot

* Country of Implementation: China
* Company: PuduTech

PuduTech is an artificial intelligence (AI) and robotics company that was founded in 2016, and it aims to design robots that perform different tasks, such as effective operation and standardized servicing and facilitating management.

Technology	Artificial intelligence (AI)/robotics technology, navigation, LiDar, ultra-wideband technology, sensors, inertial measurement unit (IMU), and encoder
Description	PuduBot is a robot designed with many technologies and sensors to be able to navigate its way through the restaurant and avoid any obstacles in its way. IT has various sensors that collect information about its surroundings which are then fused into the PuduBot's algorithms for accurate map creation; this way cm-level positioning accuracy will be achieved. Hence, this path planning will ensure that the robot takes the shortest way in turn guarantees quickness and efficiency. PuduBot is designed with trays that are placed on the inclined frame and are created using anti-slip mats that create large friction which prevents the trays from sliding or tumbling forward or backward during bumps. Using PuduBot's algorithms, the robot can perform several tasks or deliver food to multiple tables within one trip only. The workers will put orders of different tables on different trays then input the table number on the keypad found on the PuduBot. Then the robot will reach the consumers using the navigation system. The battery life of the robot is 24 hours and can hold up trays in one trip weighing 10 kg. Multiple robots can work in the same area without forming any interference problem (Brobot 2019)
Stage	In market
Beneficiaries	Consumers

Beneficiaries	Consumers	• Increased customer satisfaction • Improved customer loyalty • Faster order processing
	Stakeholders	• Multiple task performance • Increase efficiency • Saves labor costs • Improves service level • Time-efficient

Case Study 10

Developing New Flavors

- Country of implementation: USA
- Company: McCormick & Company

McCormick is a food processing company founded in 1889; it aims to manufacture, market, and distribute different flavors, spices, and seasoning mixes into the food industry. For over 125 years, McCormick was able to deliver high-quality and safe products to people around the world.

Technology	Artificial intelligence (AI)/big data
Description	Instead of relying on the human factor and the expertise of world-class chefs and flavor experts, McCormick & Company decided to delegate the task of finding its next big food flavor or new food innovation to the power of big data and artificial intelligence (AI). McCormick is developing an AI algorithm that is to be fed with huge amounts of data representing decades of market research and flavor expertise that they have accumulated over the years. The algorithm is to analyze these millions of data dots and then mimic human behavior in coming up with new flavors that should appeal to the people after looking into the huge data. This system should speed up the process of generating new flavors and spices, a process which could usually take months or years, to a great extent. It could even come up with totally new flavors that could have never be generated relying solely on the human factor. This is because the system would be unbiased and would have no personal preferences or tastes that could influence its research direction (Steffen 2019)
Stage	Prototyping
Beneficiaries	Food industry

Beneficiaries	Food industry	• Increased innovation • Expansion opportunities • Time-efficient • Cost-efficient
	Consumers	• Increased variety

Case Study 11

Recycling

| Farming | Trade | Manufacturing | Retail | Consumers |

- Countries of implementation: USA
- Company: General Mills

General Mills is a food company founded in 1928; it is one of the largest food companies in the world with operations in over 100 countries and marketing more than 100 consumer brands.

Technology	Green technology/recyclable		
Description	General Mills has been working since 2000 to reduce carbon emissions and the use of plastic in its company. To reduce the carbon footprint, General Mills' workers have made some changes to the food and the packaging. For example, the packaging team has reduced the number of pouches uploaded into each carton and the product development team has changed the shape of the pasta itself to pack the food more tightly in the package. By making these changes, General Mills was able to reduce carton size by 20% and the use of fiber paper by 890,000 lbs. as more than 99% of the fiber packaging was developed from sustainable sources. Also, they were able to eliminate greenhouse gas emissions by 11%. They also reduce the number of vehicles greatly that transfer food products; they have taken 500 trucks off the road. Moreover, General Mills uses postconsumer recycled packages to reduce the use of new plastics and reduce the carbon footprint emitted from the packaging. The company tries to get the consumers as involved as possible by creating labels called How2Recycle on the package to raise awareness and provide information about recycling. By 2030, their goal is to use 100% renewable energy (General Mills 2019)		
Stage	In market		
Beneficiaries	Environment	• Reduced carbon footprint • Reduced use of plastic • Recyclable packaging • Reduction of carton size • Environmentally friendly	
	Consumers	• Increased awareness of recycling • Environmentally friendly	

Case Study 12

Olam Farmer Information System (OFIS)

- Countries of implementation: more than 27 countries
- Company: Olam International

Olam International is the leading company in the food and agricultural industries; it was founded by Sunny Verghese in 1989. It operates in more than 70 countries and provides its food and raw materials to more than 16,200 consumers worldwide.

Technology	Internet of things (IoT)/big data, data analysis, and GPS tracking
Description	Globally cash crops like cotton, coffee beans, cashews, and cocoa are usually sourced to thousands of small farms. These farms are usually the most remote places and regular access to them is not always easy which presents a challenge connecting them to the global economy. Olam International developed the Olam Farmer Information System (OFIS) where field staff make on-ground surveys on the spot at the farms and collect all relevant data from thousands of farms including their conditions, their surrounding landscape, social circumstances, and a lot more data. This data might include farm size, location, type, and condition of crops and economic, social, and health infrastructure of the farm. This data is then inputted into the system for analysis. Data analysis later allows for giving better advice to farmers, identifying potential areas of risk such as child labor, deforestation, unexpected climate change, or low yield. Farm data and analysis enable farmers to receive more tailored support to help them improve their yields and quality, which in turn is rewarded with quality and sustainability-based premiums. Olam's sustainability partner and consumers get improved traceability and transparency with direct access to farmer and origination information. Insight derived from intervention reduces supply chain risk and improves funding efficiency (Sylvester 2019)
Stage	In market

Beneficiaries	Agriculture and farming industries	• Reduced supply chain risk • Improved funding efficiency • Better control over small factories • Higher crop yields • High-quality products • Increase in sustainability • More tailored advice and assistance • Linkage to the global economy • Reduce potential risks on the farm level
	Consumer	• Better transparency and traceability of food products to know their origins and information

Case Study 13

Reinventing Traditional Advent Christmas Calendar

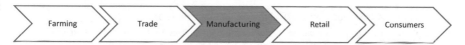

- Country of implementation: UK
- Company: Cadbury

Cadbury is a confectionery company with products that include chocolate, gum, and candy bars. It was founded by John Cadbury in 1824. It sells in more than 50 countries and includes global and local products, such as Cadbury, Clorets, etc.

Technology	Augmented reality (AR)/artificial intelligence (AI)	
Description	To engage its teenage consumers and encourage them to participate in this festive period before Christmas, Cadbury decided to utilize the AR technology where people use their smartphones to engage with the packaging or the product and get access to a wide range of interesting features. Cadbury reinvented the traditional advent calendar with the use of augmented reality on its popular "Heros" chocolate packaging. Based on the ritual of opening each advent door, teenagers could scan each Cadbury "Heros" chocolate as well as scan the AR calendar to bring the Christmas experience and feel to life. Teenagers could also earn digital currency as they take and share more selfies using the built-in AR filters and as they interact more with the app. In addition to the very high growth in sales, in this project, Cadbury succeeded in using AR as a tool to generate a huge marketing buzz as people used the filters and shared their pictures online on various social media platforms (Case study: How AR reinvented Cadbury's traditional advent calendar to drive huge engagement & social sharing 2018)	
Stage	In market	
Beneficiaries	Manufacturers	• Stronger engagement with consumers • Increased brand awareness • Marketing buzz and increased presence over social media platforms • Increased sales • Improved brand perception • Increase rates of social sharing
	Consumers	• The added value of getting entertained or acquiring useful product information using AR technology

Case Study 14

Wine Delivery Service: SIPPclub

| Farming | Trade | Manufacturing | Retail | Consumers |

- Country of implementation: UK
- Company: Sippwine

Sippwine is considered to be the best-loved wine company in the world; it was founded by Moez Seraly in 2017. It produces the best wine quality at an affordable price while at the same time guaranteeing the best experience.

Technology	Augmented reality (AR)/artificial intelligence (AI)	
Description	The sip club is an application that can be used for both IOS and Android users and is aimed particularly at millennials as they are the most wine drinkers at the moment; it reads the labels that are on the wine bottles or boxes and according to the particular type of wine, the application suggests certain meals to have with it. It also provides users with tasting tips as well as the perfect serving temperature of the particular wine. To get all the mentioned information, the user is required to point the camera at the label of the wine then all its information is displayed on the screen. This application was designed to get young people engaged with every single bottle of delivered wine. Subscribers of the application get a box with three bottles of the previously bought wine to their doorstep (Morozova 2019)	
Stage	In market	
Beneficiaries	Consumers	• Interactive and enjoyable purchasing experience • Detailed information
	Stakeholders	• Increased sales revenue • Increased customer loyalty • Increased brand awareness • Increased satisfaction

Case Study 15

Augmented Reality Food

- Countries of implementation: Global
- Company: Outreal XR

Outreal XR is a company founded in 2018 and located in Dubai, UAE; it specializes in augmented reality (AR), virtual reality (VR), and mixed reality (MR) to result in emotional engagement from consumers in different industries, including retail, education, advertising, and many more.

AUGMENTED REALITY
food

Technology	Augmented reality/artificial intelligence (AI)	
Description	This application will be able to help hotels, restaurants, and hospitality professionals widen their customer base as they will be able to promote their menu. Using the HoloMenu application, users will be able to witness a 3D experience by seeing virtual 3D food on their table when ordering online or in restaurants; they will also be able to see the different ingredients in each dish, infographics, and health and nutritional facts using different languages which enables tourists and new diners to easily be able to understand without any complications. Also, consumers will be able to view different menus and items and switch between them without the need to move to another screen. Businesses will be able to collect data about the customer's behavior, choices, preferences, etc. (HoloMenu)	
Stage	In market	
Beneficiaries	Businesses	• Engaging augmented reality experience • Increased revenues • Increased brand awareness • Brand loyalty
	Consumers	• Increase satisfaction • Informative labeling

Case Study 16

Robotic Beef Rib Cutting

Farming > Trade > Manufacturing > Retail > Consumers

- Countries of implementation: Global
- Company: Scott Automation

Scott Automation is a company that provides and designs innovative robotic solutions to increase the productivity, yield, and safety of workers. Scott Automation operates in ten countries and hence can deliver different technologies and services anywhere globally (Scott Automation).

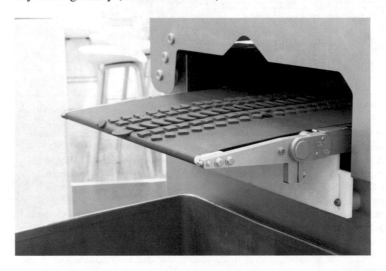

Technology	Artificial intelligence (AI)/automation, robotics technology, X-ray, 3D scanners
Description	The rib cutting process is the first process that requires a lot of skills to be performed and this is where most workers feel unsafe. Each worker requires to hold a knife in one hand and sharpening steel in the other and start cutting the bones off the meat. Using this technology, robotic beef cutting, replaces the jobs of workers that perform beef cutting operations. This robot marks the areas that need to be cut using the scribing saw and sensing technologies to accurately determine the positions that need to be cut. This technology uses X-ray and 3D scanners to present accurate and timely information or data about meat, bone, and fat consumption. Using X-ray, an anatomical representation is measured to guide the rib cutting system (Scott automation)
Stage	In market

Beneficiaries	Stakeholders	• Maximized yield • Eliminated loss of yield • Reduced health risks • Reduced labor force • Cost-effective • Increase productivity • Decreased contamination potential
	Markets	• Clean cut meat • Decreased contamination potential

Case Study 17 OAL

Eye Label and Data Code Verification: APRIL™

Farming > Trade > Manufacturing > Retail > Consumers

- Country of implementation: UK
- Company: Olympus Automation

Olympus Automation also known as OAL Group is a company that was founded in the year 1993; it aims to provide automotive manufacturing solutions in the food industry to enhance efficiency and accuracy.

Technology	Artificial intelligence (AI)/machine learning
Description	APRIL™ is the first artificial intelligence (AI) and machine learning technology that was designed for label and data code verification. This technology takes a photo or a screenshot of each data code and reads it using scanners to make sure that each one matches the organized data code. Using this technology, businesses can perform repetitive operations that require a labor force and ensure full traceability. The normal vision system used optical character recognition (OCR) and therefore struggled with packing changes, font distortion, and varying fonts and sizes. Therefore, the enhanced system uses cameras and an artificial intelligence (AI)-designed system to produce a vision system that can deal with different variations, including lighting, printing quality, and others. APRIL™ can scan 1000 packs a minute which greatly improves speed and liability (OAL group)
Stage	In market
Beneficiaries	Food industries
	• Full traceability • Reduced labor costs • Time-efficient • Eliminated error potential • Enhanced safety, quality, and efficiency
	Workers
	• Eliminated repetitive tasks

Case Study 18

Robotic Chef

> Farming ⟩ Trade ⟩ Manufacturing ⟩ Retail ⟩ Consumers

- Country of implementation: UK
- Company: Moley Robotics

Moley Robotics is a company that produces and develops robotic machines that can perform tasks in the kitchen. It was founded by Mark Oleynik in 2015.

Technology	Artificial intelligence (AI)/robotics technology and automation	
Description	The robotic chef is designed using robotic hands that consist of 20 motors, 24 joints, and 129 sensors to create the exact movements of a professional chef. It can operate with the same speed, movement, and sensitivity. The movement of the robotic hands is based on the cooking skills of Master Chef Tim Anderson who is the winner of the BBC MasterChef; all his motions and cooking skills were recorded and then the company was able to design a series of algorithms on the system of the robotic chef. The robotic chef can cook any recipe even a creation of the user's design. These robots will soon be able to operate in people's homes. The robotic arms are paired with an oven, stove, dishwasher, fridge, and small appliances to make it suitable to do all different tasks of a kitchen. The designed system is operational using the touchscreen unit or an application that is developed to enable remote access from smartphones. The entire system of the robotic arms with its kitchen appliances is built behind sliding glass to maximize safety at all times (Williams 2020)	
Stage	In market	
Beneficiaries	End users	• Cleaning, cooking, measuring, dishwashing, etc. • Increased accuracy of seasoning and ingredients • Unlimited access to recipes around the world • Safe • Distance operated • High-quality dishes prepared quickly and efficiently

Case Study 19

SmartLabel

- Countries of implementation: Global
- Company: Trading Partner Alliance

Trading Partner Alliance is a company founded in 2001 that produces innovative ways and solutions to ensure intelligent communication between consumers and businesses.

Technology	Blockchain/software
Description	SmartLabel technology can provide consumers with all the information necessary about any product they want. It gives instantaneous information about thousands of food products and beverages. They can trace the product back to its origin and see all the steps in its production. They are also able to view details about its nutrition, ingredients, allergens, sustainability, company, calories, and more. Consumers can trace information from the smartphone after scanning the label on the product. At the moment, SmartLabel can view more than 9000 products and 500 brands (SmartLabel)
Stage	In market

Beneficiaries	Consumers	• Detailed product information • QR access product information • Traceability of sourcing and distribution • Increased customer satisfaction • Informed decision-making
	Brands	• Increase brand awareness • Improved customer loyalty

Case Study 20

Pizza Manufacturing

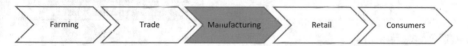

- Country of implementation: USA
- Company: Grote Company

Grote Company is a privately held company that was founded in 1972; it designs machines to enhance food slicing and assembling. Their technologies are used to make pizzas, sandwiches, etc.

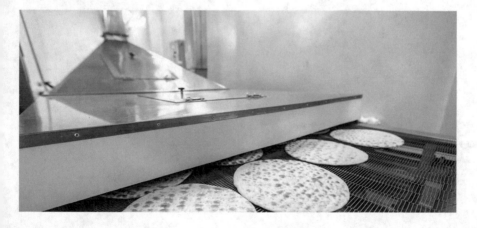

Technology	Artificial intelligence (AI)/robotics technologies and automation

Description	Grote company designed a machine responsible for the pizza topping production line. It automates frozen pizza by having a sauce applicator, topping applicator, and slicing equipment all in one machine; this machine is available to operate in a single lane or multiple lanes. It can operate on 45 pizzas per minute and applies the different topping based on the order of the production needs. Using border-free and waterfall applicators, the machine can meet the needs of the producer and add the ingredients accurately and uniformly onto the pizzas. The sauce applicators add sauce patterns accurately onto the flat and raised edge pizza style. The topping applicators apply a great variety of toppings, including shredded cheese, dry ingredients, and meats or vegetables. This technology can handle different crust sizes, from 5″ up to family size (Grote company)	
Stage	In market	
Beneficiaries	Stakeholders	• Fast, sanitary production • Increased efficiency and production • Increased yield • Reduced waste • Elimination of potential errors • Personalized specifications

Health Care

Abstract Health care is a broad industry whose primary purpose is to provide medical care or assistance to a society or a community; it involves improving someone's physical or mental health by medical professionals who guide their patients through the right recovery journey. The request for health care is increasing and changing in response to the rise of new diseases and chronic conditions. There are several stakeholders in the health-care system, and each of them has different goals to achieve in health care. The primary stakeholders found in this industry are patients, insurance companies, physicians, and pharmaceutical firms.

Keywords Disease detection · AI diagnosis · Handwriting recognition · Predictive data · Prediction through machine learning · Wearables · Health monitoring · VR technology · Bioprinting · Digital dentistry · Robotic surgery system · Electronic health care · Microorganisms · Nanotechnology · Surgical systems · AI medical consultation · Digital nurse

Health care is a broad industry whose primary purpose is to provide medical care or assistance to a society or a community; it involves improving someone's physical or mental health by medical professionals who guide their patients through the right recovery journey. The request for health care is increasing and changing in response to the rise of new diseases and chronic conditions. There are several stakeholders in the health-care system, and each of them has different goals to achieve in health care. The primary stakeholders found in this industry are patients, insurance companies, physicians, and pharmaceutical firms.

Health care has improved drastically over the past 100 years, where there have been vast levels of innovation advancement, such as X-rays, organ transplants, MRI, anesthetic machines, and many more. One of the significant changes that this industry is trying to achieve is the digital transformation in health, which is the use of different technologies to help monitor someone's health; it covers everything from sensors to artificial intelligence (AI) to electronic records. Digitalization in health care has risen the opportunity to prevent diseases, improve diagnosis, monitor patients, and lower health-care costs. With digital health tools, professionals could monitor their patients and diagnose early symptoms, which will help shorten the length of disease and improve quality of life (The digitally Engaged Patient: Self-Monitoring and Self-Care in the Digital Health Era I Springlink 2013).

As health care is continuing to evolve every day, there are still numerous challenges that keep arising. One of the challenges that health care faces today is security; with so many health programs and data reading programs on mobile phones, smartwatches, etc., there is a considerable threat of data breaches and security problems. The more complex the technologies used in health care, the higher the risk of the data being hacked (Security and Privacy Issues with Healthcare Information Technology I IEEE Xplore 2006). Another challenge is the medical error, which can occur due to poor diagnosis of a physician when not conducting the right or few examination tests or when there are technical failures, which are the complications that can occur in medical equipment. On the bright side, the opportunities for health-care success nowadays are increasing significantly; each day, new ways of using different technologies are rising. Hence, health-care engineering is a big part of the health-care industry today as it is constantly introducing new technological innovations.

Using such technologies will help the doctor read data analysis about the patient's health and predict the patient's needs. For example, artificial intelligence (AI) will help doctors access information faster and deliver a better and more accurate diagnosis.

Value Chain

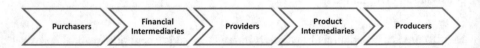

The value chain of the health-care industry is comprised of five key players: the purchasers, financial intermediaries, service providers, product intermediaries, and producers. The purchasers are those who pay for the health-care services which could include individuals, their employers, or government organizations. Financial intermediaries refer to insurers. The service providers are those that deliver health-care services such as clinics, hospitals, pharmacies, and health-care networks and directories. Product intermediaries include wholesalers and retailers of medical equipment and finally the producers of such equipment.

Case Studies

Case Study 1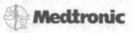

GI Genius Intelligent Endoscopy Module

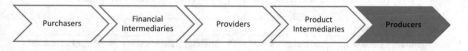

Purchasers ⟩ Financial Intermediaries ⟩ Providers ⟩ Product Intermediaries ⟩ **Producers**

- Country of implementation: Ireland
- Company: Medtronic

Medtronic is a medical device company that develops technologies, such as drug delivery devices, mechanical devices, and surgical instruments. It was founded in 1949; however, recently, the company decided to use artificial intelligence (AI). It employs more than 90,000 people worldwide, and it serves hospitals, doctors, and patients in 150 countries (Medtronic Launches the First Artificial Intelligence (AI) System for Colonoscopy at United European Gastroenterology Week 2019. | Medtronic 2019).

Technology	Artificial intelligence (AI) /data analytics	
Description	Medtronic designed a product called GI Genius, a module that uses artificial intelligence (AI) to detect and prevent colorectal cancer. It highlights the regions with different types of mucosal (the inner lining of some organs and body cavities) abnormalities. The GI Genius is the first module worldwide to use artificial intelligence (AI) to observe colorectal cancer. It detects early symptoms that may go undetected, and it is incredibly accurate in finding lacerations on the mucosa, hence increasing accuracy (Medtronic Launches GI Genius | PharmaTimes 2019)	
Stage	Recent to market	
Beneficiaries	Health-care professionals	• Prescribing patients with the correct medicine • Providing the right treatment • Detecting early symptoms of colorectal cancer
	Patients	• Ability to accurately detect cancer at an early stage, which will help the patient start treatment before the cancer spreads, minimizing damage • Less invasive treatments • Cost-effective • Time-effective

Case Study 2

Quicker Diagnosis Using Machine Learning

- Country of implementation: USA
- Company: PathAI

PathAI is a company that develops technologies to assist pathologists in making a better and more rapid diagnosis. It was founded in 2016 and aims through technology to enhance patients' diagnosis in health care (PathAI Case Study \| Aptible).

Technology	Artificial intelligence (AI)/machine learning
Description	PathAI has created an AI pathology and diagnostic tool which uses the power of machine learning by teaching the computer to detect patterns in previous diagnoses and treatments. Therefore, they help pathologists to make a more accurate diagnosis in a shorter amount of time. They also can use the very same method to identify patients who might respond to and benefit from a specific type of therapy (15 Examples of Machine Learning in Healthcare That Are Revolutionizing Medicine | Builtin 2020)
Stage	In market
Beneficiaries	Professionals

	Professionals	<ul><li>The large amount of funding to support AI algorithms for better diagnosis</li><li>Better and more accurate diagnosis</li><li>Minimizes tedious work and reduces time to diagnose a patient</li></ul>
	Patients	<ul><li>Rapid and accurate diagnosis</li><li>Recommendation of therapy sessions that will ease their recovery journey</li><li>New and better treatments</li></ul>

Case Study 3 **CiOX** HEALTH

Enhancing Information Management Using ML

Purchasers — Financial Intermediaries — Providers — Product Intermediaries — **Producers**

- Country of implementation: USA
- Company: Ciox Health

Ciox Health is a health-care information management company that manages, records, and shares health information within the health-care industry with high data security. Their main goal is to provide the health-care industry with advanced methods efficiently to exchange and manage health information (About Ciox | Ciox).

Technology	Artificial intelligence (AI/machine learning)	
Description	Ciox has created a new machine learning platform, called HealthSource, to overcome challenges in health-care data management. It was developed to help clients to share and manage different types of health-related data. HealthSource assists hospitals to manage and reach health information seamlessly. The platform has additional technologies such as handwriting recognition and neural network tools to convert unstructured data into a digital form (New Machine Learning Platform From Ciox Applies AI to Interoperability | Healthcare IT News 2018)	
Stage	In market	
Beneficiaries	Health-care professionals	• Provide clients with better health-care data improving the accuracy and the flow of medical records • Increase efficiency and transparency to the release of memorial information • Improve accuracy • Easier access to health information
	Patients	• Patients who use any health-care facility that has improved by the availability of these clinical data are benefiting indirectly by receiving better health care

Case Study 4

ML in Understanding the Body's Immune System

| Purchasers | Financial Intermediaries | Providers | Product Intermediaries | **Producers** |

- Country of implementation: USA
- Company: Pfizer

Pfizer is a multinational innovative pharmaceutical corporation. The company developed and created hundreds of different vaccines and medicines worldwide. It is one of the leading research-based pharmacy institutions that aim to provide the public with better health care and effective medications to improve their health (About Us I Pfizer).

Technology	Artificial intelligence (AI)/machine learning
Description	Pfizer collaborated with IBM to use its developed Watson AI technology to answer questions formulated in a natural language in health care. Pfizer has used its machine learning technology in its immune-oncology research to learn how the body's immune system can fight cancer. Pfizer uses Watson to analyze various data and produce insights based on evidence and tested theories. Furthermore, Pfizer extended its coordinated efforts with a Chinese tech startup XtalPi to build up an artificial intelligence (AI) – a powered platform to model small-molecule drugs that shall combine quantum mechanics along with machine learning to predict the properties of many molecular compounds (15 Examples of Machine Learning in Healthcare That Is Revolutionizing Medicine I Builtin 2020)

Stage	Recent to market	
Beneficiaries	Health-care industry	• New drug discoveries
	Patients	• Creation of advanced drugs to fight cancer

Case Study 5

Internet of Things (Internet of Things (IoT)) Devices for Diabetes Monitoring

| Purchasers | Financial Intermediaries | Providers | Product Intermediaries | **Producers** |

- Country of implementation: UK
- Company: National Health Service (NHS)

NHS incorporates all health-care systems that are funded by the public in the UK. It is one of the largest health-care systems in the world and is mainly financed by the government. The objective of this system is to provide UK residents with free health-care services (National Health Service | Wikipedia).

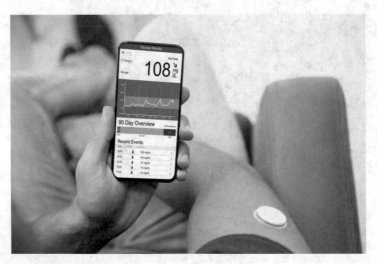

Technology	Internet of things (IoT)	
Description	On the World Diabetes Day of 2018, NHS started producing continuous glucose monitors (CGMs) which are devices that use multiple sensors to help diabetics continuously monitor their blood glucose level for 7 days by taking different readings at regular intervals. Afterward, they launched different mobile apps to allow the wearer of CGM devices to check their readings and detect trends quickly (10 Examples of the Internet of Things in Healthcare I Econsultancy 2019)	
Stage	In market	
Beneficiaries	Caregivers	• One of the apps is "The FreeStyle LibreLink" app, which allows the remote caregiver to monitor their patients whether they are the parents of their diabetic children, diabetic relatives, or clinical patients
	Patients	• Quick and easy way to monitor their condition and treatment • A noninvasive, convenient way to know a patient's glucose level

American Society o
Clinical Oncology

Case Study 6

Internet of Things (IoT) in Cancer Treatment

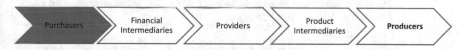

Purchasers → Financial Intermediaries → Providers → Product Intermediaries → Producers

• Country of implementation: USA
• Company: American Society of Clinical Oncology

ASCO provides the highest quality resources for cancer care. The company's main objective is to help cure cancer through advanced research and high-quality education to improve patient care (About ASCO I Cancer.Net).

Technology	Internet of things (IoT)	
Description	To fulfill ASCO's main going of finding a cure for cancer, the company conducted a clinical trial using the CYCORE system. The system consists of sensors, Bluetooth technology to allow blood pressure monitoring and weight scaling all under a mobile app. This app monitors patients' symptoms and treatment outcomes daily. The purpose of this trial is to see if the potential of using smart technologies in cancer treatment would be beneficial. The results were that patients monitored under the smart monitoring system, CYCORE, were showing better results than those who continued to go to their physician weekly without extra smart monitoring (10 Examples of the Internet of Things in Healthcare | Econsultancy 2019)	
Stage	Recent to market	
Beneficiaries	Physicians	• It has improved the contact between the physicians and the patients as they monitor their condition with minimal interference in patients' daily lives
	Cancer patients	• They experience less harsh symptoms as their physicians are continuously following remotely along with them • Remote monitoring that allowed people not to be tied down to their houses and reduced hospital visits

embodied
labs

Case Study 7

VR Simulations to Better Understand Real Patients

- Country of implementation: USA
- Company: Embodied Labs

Embodied Labs allow users to experience what it is like to live with the symptoms of a certain health condition. It uses interactive 360-degree videos and VR technologies to embody what people suffer from (Embodied Labs | LinkedIn).

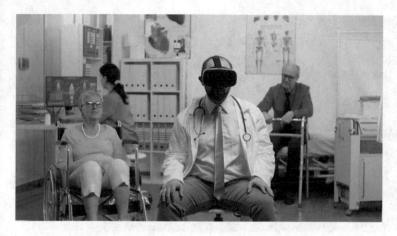

Technology	Virtual reality/VR and 360 videos	
Description	The company uses virtual reality technology along with 360 videos to simulate the experiences and feelings of real patients. "We Are Alfred" is one of the company's first labs which transforms the experience of the 74-year-old African American Alfred with high-frequency hearing loss and macular degeneration through a VR headset. They can also control spatial sound by adjusting what people can hear if they lean toward the source of the noise from the video (Can Virtual Reality Change The Way We Think About Health? | Builtin 2019)	
Stage	In market	
Beneficiaries	Caregivers	• Understanding of real symptoms of patients by virtually experiencing them enables caregivers to provide better treatment and care
	Patient	• Patients receive better care and treatment

Case Study 8

Augmented Reality (AR) for Medical Records

- Country of implementation: USA
- Company: Augmedix

Augmedix's mission is to rehumanize the relationship between the doctor and the patient by addressing the problem of documentation via providing the live clinical report (About Us | Augmedix).

Technology	Virtual reality	
Description	Augmedix uses Google Glasses to help physicians in the USA to remotely document patient visits to create a more natural doctor-patient relationship. The technology gives the doctor quick access to all these medical records facilitating a more natural interaction. This method helps reduce the time and effort the physician puts into tedious paperwork. Thus, the Augmedix platform enables doctors to see more patients and have accurate documentation of their visits. According to Augmedix, this new technology helped physicians to see up to 25% more patients (Augmedix: Humanizing Healthcare Through Google Glass | Digital Initiative)	
Stage	Recent to market	
Beneficiaries	Physicians	• Doctors have quick and efficient access to their patient medical records on the spot • Accurate documentation to visits
	Patients	• More access to physicians • Better health-care experience

Case Study 9

Using 3D Printing in Brain Research

Purchasers	Financial Intermediaries	Providers	Product Intermediaries	Producers

- Country of implementation: USA
- Company: Wake Forest Institute for Regenerative Medicine

WFIRM is a company that specializes in research and developing medicine to cure diseased tissues and organs to restore their normal functions. The company uses technological advancements to provide solutions for medical challenges. Some projects include developing cells that produce insulin and blood vessels for heart surgeries (Wikipedia | Wake Forest Institute for Regenerative Medicine).

Technology	3D printing (additive manufacturing)	
Description	One of the techniques used in 3D printing is bioprinting, which involves printing layers of living cells instead of plastic or metal. It uses a computer-guided pipette known as the bio-ink to build artificial living tissues in laboratories. These organs and tissue constructs resemble our real body organs and tissues, allowing us to perform research on them on a miniature scale. They also allow users to use them as a cheaper alternative for organ transplants as they function in the same manner as the natural organs. Wake Forest Institute for Regenerative Medicine has built 3D brain organoids with a lot of potential for drug discovery and disease modeling. They announced in 2018 that their model is full cell-based and functional like the normal brain anatomy (3D Printing in The Medical Field: Four Major Applications Revolutionizing The Industry I Verdict Medical Device 2018)	
Stage	Recent to Market	
Beneficiaries	Researchers	• The mimicking and modeling of human anatomy may lead to drug discovery • Removed the need for humans for an experimental drug trial

Case Study 10 ᗕᐴENVISIONTEC

3D Printing Dental Parts

| Purchasers | Financial Intermediaries | Providers | Product Intermediaries | Producers |

- Country of implementation: USA
- Company: EnvisionTEC

EnvisionTEC is an international company founded in 2002 that manufactures and sells various 3D printer-based products from digital design files (About EnvisionTEC I EnvisionTEC).

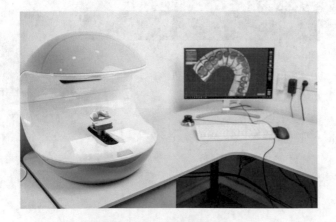

Technology	3D printing	
Description	Digital dentistry has become more popular, and EnvisionTEC is contributing to its popularity. EnvisionTEC is using 3D printing technologies to help the density industry. The company provides high-quality, low-cost, accurate, and full-production 3D printers to produce crowns and orthodontic models with a smooth, clean finish. They also built an industry-leading materials library of FDA-approved materials for their machines. They design their materials such that they are flexible and deliver reliable results. The materials the company produced for its 3D printing allows the dentist to build denture bases, surgical dill guides, crowns, bridges, dental frameworks, and a lot more applications (EnvisionTEC Dental 3D Printers: Accurate, Fast, Reliable, And Flexible I EnvisionTEC)	
Stage	In market	
Beneficiaries	Clinicians	• Cost-effective • Time-efficient • Increased control of treatment plans
	Dental Labs	• Increased product competitiveness

Case Study 11  ACCURAY®

Using Automation in Tumors Therapy

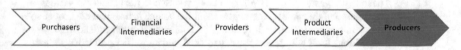

- Country of implementation: USA
- Company: Accuray Incorporated

Accuray is a radiation oncology company that develops innovative tumor therapies to help people live a long healthy life. The company mainly constructs devices that are used in the health-care industry as radiation therapy for cancer patients (Accuray Incorporated I Reuters).

Technology	Automation and digitalization
Description	Accuray has developed the CyberKnife robotic surgery system that provides radiation therapy to tumors with a precision of a fraction of millimeters. It has been first invented in the 1990s, and now it is used to treat cancer at various hospitals and treatment centers. The radiation source is installed on a robot so that it allows for targeting the beam of radiotherapy to a tumor. The laser repositions itself with high accuracy to reach the tumor from all sides without repositioning the patient. It also provides therapy for some part of the body that was hard to reach surgically like the prostate, head, and neck (5 Medical Robots Making a Difference in Healthcare | Case Western Reserve University 2017)
Stage	In market
Beneficiaries	Patients

Beneficiaries	Patients	• Convenient, noninvasive tumor removal • Minimized damage to healthy tissue due to high precision
	Surgeons	• Removing tumors accurately, more effectively, while minimizing damage to healthy tissue

Case Study 12

Using Digitalized Records in Hospitals

| Purchasers | Financial Intermediaries | Providers | Product Intermediaries | **Producers** |

- Country of implementation: USA
- Hospital Name: Kimmel Pavilion

Kimmel Pavilion is one of NYU Langone's in-patient care facilities that create healing the atmosphere for children and adults by including integrated technologies. It was the first hospital to provide private rooms for patients in New York with a new advanced environment that enabled an overall better health-care experience (NYU Langone Health, Helen L. and Martin S. Kimmel Pavilion | Ennead).

Technology	Automation and digitalization/sensors	
Description	The hospital's IT department has integrated the electronic health-care record with another health-care workflow application. Therefore, if, for example, a patient is leaving and discharge, his electronic health record will immediately notify the workflow record that his room is empty now and that it needs cleaning. Hence, it schedules a housekeeping appointment to clean it (Innovative Hospitals Tap Automation to Streamline Patient Care | Health Tech 2018)	
Stage	In market	
Beneficiaries	Hospital's operations	• Enhanced efficiency and optimized patient care • Time-effective • Cost-effective

Case Study 13 DISINFECTION SERVICES™

Using Automated Robots in Disinfection Processes

Purchasers	Financial Intermediaries	Providers	Product Intermediaries	Producers

- Country of implementation: USA
- Company: XENEX Disinfection Services

XENEX is a privately held company that specializes in disinfecting health-care facilities. The company's objective is to help enhance patient's health-care experience by removing any harmful microorganisms that may lead to the rise of the clinic obtaining infections (RevDesinfectie Robots Deploys Xenex LightStrike Robots for Contamination Control in Pharmaceutical Cleanrooms I Business Wire).

Technology	Automation and digitalization/automated robots
Description	Health-care facilities struggle to clean and thoroughly sanitize their spaces between patients because of the limited time and the difficulty of seeing and destroying microorganisms. A lot of patients who are ill and immunocompromised are very vulnerable to bacterial disinfection. Therefore, XENEX decided to use an automated and portable robot that disinfects hospital rooms in a matter of minutes. This is achieved by using pulsed full-spectrum UV rays that kill different types of infectious bacteria (5 Medical Robots Making a Difference in Healthcare I Case Western Reserve University 2017)
Stage	In market

Beneficiaries	Health-care facilities cleaners	• Increased efficiency and productivity in disinfecting rooms • Reduced time for disinfection
	Patients	• Minimizing infections • Safer environment

Case Study 14

Using Nanotechnology in Pharmaceuticals

Purchasers → Financial Intermediaries → Providers → Product Intermediaries → Producers

- Country of implementation: USA
- Company: Amgen Inc

Amgen Inc. is an American multinational pharmaceutical company based in California, and it includes a large number of scientists focusing on molecular biology and biochemistry. The company's goal is to develop medicines that help destroy dangerous illnesses through innovative ideas (About Amgen | Amgen).

Technology	Nanotechnology
Description	Amgen uses nanotechnology to develop medicines for severe illnesses. Their nanotechnology platform is used in medical imaging, diagnosis, and delivery of drugs, all of which will help treat those severe diseases. This may enable medicines to have better outcomes by reducing their toxicity and having a more effective delivery of drug molecules into the body (Nanotechnology in Medicine: Who Are the Leading Public Companies I Verdict Medical Device 2019)
Stage	Testing phase
Beneficiaries	Patients	• Minimized toxicity in drugs • Efficient delivery of drug molecules • Longer life span of medicines in the body

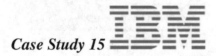

Case Study 15

Artificial Programs Pinpoint Cancer Treatments

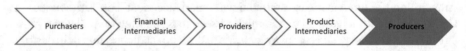

Purchasers → Financial Intermediaries → Providers → Product Intermediaries → **Producers**

- Country of implementation: USA
- Company: IBM

IBM is an American international company that is based in New York and specializes in technology. It develops and manufactures computer hardware, middleware, and software devices and also offers technology consulting services (IBM I Wikipedia).

Technology	Artificial intelligence (AI)/machine learning and data analytics	
Description	IBM created an AI platform called IBM Watson Health used in the health-care industry to provide professionals and researchers with actionable insights after analyzing a vast amount of data to make a more informed decision on the course of treatment. This method is being used to overcome one of the main challenges the health-care sector face, cancer treatments. The platform analyzes a large amount of data, which is too complicated for doctors to study, inspect, and present a customized treatment for the patient (The Changing Oncology Landscape I IBM)	
Stage	Recent to market	
Beneficiaries	Physicians	• The faster, more efficient way to analyze complex and extensive data • Save time and effort to run through unstructured data • Provide a better and accurate insight to course of treatment
	Patients	• Customized, more accurate treatment • Faster diagnosis

Case Study 16 INTUİTIVE

Robots Making a Difference in Health Care

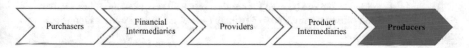

- Countries of implementation: Global
- Company: Intuitive Surgical

Intuitive Surgical is an American company specializing in developing, creating, building, and advertising robotic technologies that aim to increase efficiency and productivity within the health-care industry. The company's purpose in manufacturing those robots is to create a less invasive and sophisticated approach to surgeries (About Intuitive I Intuitive).

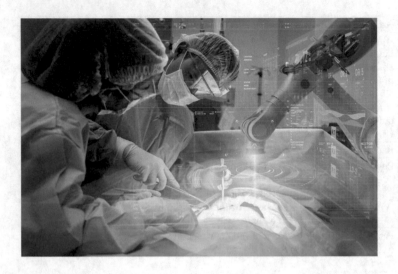

Technology	Automation and digitalization/robotics	
Description	The da Vinci Surgical System is a robotic surgical tool that aids surgeons in conducting surgeries with minimum invasion, thus reducing medical errors. The system has developed innovative tool and a high-definition 3D view of the surgical section. This system has been installed in thousands of hospitals worldwide and can be used for surgical procedures consisting of heart valve repair, prostatectomy, and gynecology (About Da Vinci Systems I Intuitive)	
Stage	In market	
Beneficiaries	Health-care professionals	• Better surgical precision • Has a wider range of motion • Enhanced visualization with the high-definition 3D camera • Easier access to areas that are hard to reach
	Patients	• Higher chance of surgical success • Minimizes risk of medical error and infection • (Robotic Surgery I Cancer Treatment Centers of America)

Case Study 17

Digital Consultation

| Purchasers | Financial Intermediaries | Providers | Product Intermediaries | **Producers** |

- Countries of implementation: UK, Rwanda, and Canada
- Company: Babylon Health

Babylon Health is a company that provides health-care services remotely where professionals in the medical industry may provide advice with text or video calls. It offers its services through its website or downloadable mobile apps. The company aims to provide low-cost, attainable medical services with the use of artificial intelligence (AI) (Babylon Health is Building an Integrated, AI-Based Health App to Serve a City of 300k in England | Tech Crunch 2020).

Technology	Artificial intelligence (AI)/mobile application and digitalization
Description	The Babylon Company has created an AI-based medical consultation application called the Babylon App. The form offers actions and recommendable insights based on the user's medical history, known medical information, and a database on different illnesses. The platform uses speech recognition where clients can "Ask Babylon" questions about any medical concerns. Additionally, patients may access their medical records, book appointments, and renew a prescription (The Amazing Ways Babylon Health is Using Artificial Intelligence (AI) to Make Healthcare Universally Accessible	Forbes 2019)
Stage	In market

Beneficiaries	Patients	• 24/7 access to health-care professionals • Easy, more convenient, affordable access to medical consultations • No confirmation bias

Case Study 18

Personalized Prosthetics

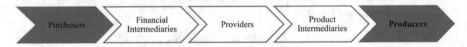

| Purchasers | Financial Intermediaries | Providers | Product Intermediaries | Producers |

- Countries of implementation: Global
- Company: Ekso Bionics

- Ekso Bionics is a company that builds and creates bionic exoskeleton systems that are wearable. The exoskeleton robotic device aims to improve robustness, mobility, and tolerance in rehabilitation and across the health-care industry. The company aims to enhance natural abilities and enhance the quality of life (About Us | Ekso Bionics).

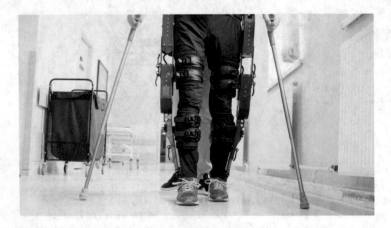

Technology	3D printing/robotics
Description	Ekso Bionics manufactured a robotic exoskeleton device called EksoGT that helps people with motor impairment, especially in lower extremities, to abandon their wheelchairs and walk again. The device can enhance and substitute human abilities. It is developed for the medical center to give victims of spinal cord injuries or stroke patients computer guidance to get back on their feet (Worldwide Adoption of Ekso Bionics EksoGT Exoskeleton Allows Stroke and Spinal Cord Injury Patients to Take 100 Million Steps to Date | Globe Newswire 2018)
Stage	In market
Beneficiaries	Patients	• Helps patients to walk again • Mobilizes patients faster, easier, and more effectively

Case Study 19 SENSELY

Digital Nurse

- Countries of implementation: Global
- Company: Sense.ly

- Sense.ly is an American company that develops avatars and chatbot programs to use in the medical industry. It aims to provide practical help for patients or clients by integrating responsive conversation and technological advancements. It offers technical virtual-based, actionable insights to clients, hospitals, and insurance companies (Sense.ly Overview).

Technology	Augmented reality (AR)/platform and machine learning
Description	Sense.ly has created a digital nurse called Molly using AI technology that supports and updates patients' conditions and prescriptions. Patients check-in with Molly through a mobile app that uses speech recognition to share medical information that professionals can review. To make the app more flexible, Sense.ly incorporated in its algorithms the popular medical protocols and diagnoses, especially for those who suffer from chronic diseases (Virtual Nurse App Sense.ly Raises $8 Million From Investors Including the Mayo Clinic I Techcrunch 2017)
Stage	In market
Beneficiaries	Patients • Ability to provide virtual medical attention • Long-term, personalized monitoring, and care

Case Study 20 Microsoft

Virtual Surgical Intelligence

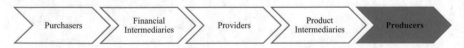

- Countries of implementation: Global
- Company: Microsoft

Microsoft is an American corporation that specializes in computer technology. It is the largest and most significant software company worldwide. It develops, builds, and sells software, electronics, computers, and any computer-related services. The company aims to create top-notch platforms and software that help empower people and different organizations (A Brief History of Microsoft - The World's Biggest Software Company | DSP).

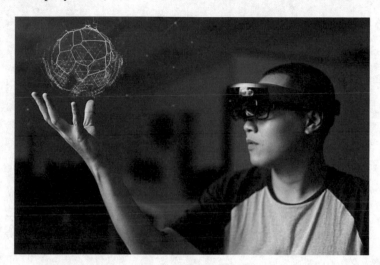

Technology	Augmented reality (AR)/virtual reality	
Description	Microsoft is changing medicine and surgery by using its revolutionary HoloLens technology in the health-care industry. One of the medical applications the HoloLens does is presenting a 3D model of human anatomy to users where users can go through different layers of skin, muscles, and organs. With this technology, it became possible to realistically simulate and experience any illness or conditions such as heart problems or damaged lungs. It is possible to model the human body aspect with detail while having all possible points of view. This technology could also be used to preplan or practice surgical operations. Surgeons could accurately make small incisions and vividly see the results of their actions (How Microsoft's HoloLens is Changing Medicine and Surgery | Futurum 2017)	
Stage	In market	
Beneficiaries	Healthcare Professionals & scholars	Ability to have virtual education instead of physical dissection Learn about anatomy more effectively Having a more vivid representation of the human body (better than books) Fruitful learning by turning 2D images into 3D augmented models More effective and better training and education (Mixed Reality in Healthcare - The HoloLens Review	The Medical Futurist 2017

Media and Entertainment

Abstract The media and entertainment industry is one of the most significant global sectors and one of the first adopters of innovative technological advancements. The industry is divided into segments, such as film, TV shows, radio shows, music, newspaper, magazines, and books. Its strategies are continually evolving as technology advances. The media industry plays a crucial role in influencing and constructing the public's opinion while the entertainment sector brings people together and helps create a sense of relaxation and happiness (How is technology Transforming the Media and Entertainment business | Jellyfish Technologies).

Keywords Digital music · Interactive movies · AR games · Streaming · Immersive Audi · Wearables · Noise cancelling · High-resolution imaging · Nano coloring · 3D mapping · Automated reports · Real-life experiences · Customer engagement · Data analysis · Consumer analytics · AI director

The media and entertainment industry is one of the most significant global sectors and one of the first adopters of innovative technological advancements. The industry is divided into segments, such as film, TV shows, radio shows, music, newspaper, magazines, and books. Its strategies are continually evolving as technology advances. The media industry plays a crucial role in influencing and constructing the public's opinion while the entertainment sector brings people together and helps create a sense of relaxation and happiness (How is technology Transforming the Media and Entertainment business | Jellyfish Technologies).

As the internet and mobile technology increase in popularity, the advertising business revenues are susceptible to consumer behavior to follow, subscribe, and like the content. Today, digitalization dominates the growth of the media and entertainment industry. Disney media networks were impacted as consumers shifted to watching online content on platforms like YouTube, Netflix, and other streaming services. Another major issue the industry faces is piracy and cyber risks. It has become imperative for this industry to protect its consumers and their content (10 Challenges and Opportunities Media and Entertainment Industry in 2018 | LinkedIn 2018).

© Springer Nature Switzerland AG 2023
M. Anis et al., *Mapping Innovation*,
https://doi.org/10.1007/978-3-030-93627-3_8

Business costs are escalating as media research to stay up to date with evolving market trends and technology is constantly needed. Another challenge the industry faces is the craving for innovation and creativity. The industry needs to prioritize creating content that is appealing with a digital experience that connects with consumers in a meaningful way.

During the COVID-19 pandemic, the media and entertainment industry was severely affected with the lockdown, live performances, outdoor recreational activities, and cultural site visits being postponed or canceled. The pandemic leads to the global shutdown of TV and film production as well as cinemas, live entertainment venues, and theme parks. Since the media and entertainment industry largely depends on social activity, the current world's circumstance was the most significant challenge this industry had to overcome (Media and Advertising are Changing. Here are your Biggest Challenges. | Inc 2019).

There remain however many opportunities in the media and entertainment industry. With digitalization becoming inevitable, data on consumer viewing and shopping behavior interests, attitude, and demographics are easily attained by modern analytical applications. The use of targeted data can allow companies to deliver precisely targeted content or advertisements. Additionally, the industry's shift to online data has become a game changer. There are opportunities for new talents from professionals to produce creative content and creative client servings. Companies now have more opportunities to create virtual entertainment and real-life modeling activities, virtual events, and virtual museums. The industry is growing and implementing innovative technologies to create appealing and personalized content and build a high-quality online experience (5 Benefits: Competitive Advantages of big data in Business | new-gen apps 2017).

Cloud Collaboration enables multiple collaboration remotely and simultaneously on content creation projects through dedicated software. A band can record virtually, in real-time with the same recording quality regardless of each member's location. Through high-speed internet, the latest software can mirror recording sessions, at several locations, without identifiable delays ensuring recording continuous around seamlessly around the globe. This technology introduced in AVID Pro Tools, a digital audio workstation in 2018, is stored in the Cloud, and it can be accessed from different locations. Users can manage different versions of a project and give reviews independently, similar to Google's docs and sheets (Cloud Collaboration | SearchCloudComputing 2015).

Value Chain

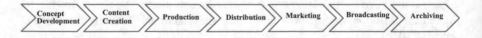

The value chain for media and entertainment value chain comprises seven phases: concept development, content creation, production, distribution, marketing, broadcasting, and archiving.

Concept development involves the identification of appropriate content to be created and delivered through understanding the needs and trends of the market. Content creation involves the production of relevant materials, through activities, such as creating new shows, written content, or visuals. In the production phase, created content is fully developed, ready for consumption. Products are then ready for distribution by selling to outlets such as TV networks, cinemas, bookstores, etc. Once there the content gets promoted through various marketing activities, before being broadcast. Once the broadcasting phase has been concluded, content is archived for future use, if necessary. This archiving phase concludes the value chain for this industry.

Case Studies

Case Study 1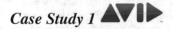

Online Co-creation and File Sharing Using Cloud Collaboration Technology

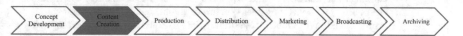

- Countries of implementation: Global
- Company: AVID Technology

AVID Technology is an American technology and multimedia company that has a good reputation in the media and entertainment industry. It specializes in audio and video, specifically, digital nonlinear editing (NLE) systems, video editing software, audio editing software, music notation software, management, and distribution services. Avid products are used in the television and video industry to create television shows, feature films, and commercials. Avid products are used in the television and video industry to create television shows, feature films, and commercials (Avid Acquires the Assets of Rocket Networks| businesswise 2003).

Technology	Platforms/cloud collaboration
Description	Pro Tools is one of the very few technologies that dominate this industry. It was developed to create one of the best quality digital music recordings in the industry. Pro Tools was created by Avid technology, and its purpose is to provide a digital audio workstation to develop one of the world's most diverse sets of recording, editing, and mixing system. Pro Tools' software enables musicians to record using analog hardware, edit and mix songs with the built-in editing and mixing tools, add effects, and save them on a CD. The technology's hardware is responsible for converting analog audio to digital signals the computer can comprehend. Avid Cloud Collaboration for Pro Tools is a feature developed in 2018 to enable multiple collaborators to contribute remotely and simultaneously on the same project (Pro Tools Systems I HowStuffWorks).
Stage	In market

Beneficiaries	Recording artists, musicians, sound engineers, postproduction artists	• Remote access • Convenience • Improved health and safety • Cost-effective • Collaboration • Cloud storage • (Advantages and Disadvantages of Pro tools I HowStuffWorks)

Case Study 2

Online Streaming and Smart Media

Concept Development	Content Creation	Production	Distribution	Marketing	Broadcasting	Archiving

- Countries of implementation: Global
- Company: Netflix

Netflix is the world's leading streaming entertainment company that allows users to watch a diverse range of TV shows, movies, documentaries on any internet-connected device.

Technology	Artificial intelligence (AI)/cloud-based tools		
Description	Netflix came up with a movie as part of the Black Mirror series called Bandersnatch. It is an interactive movie where the watchers' decisions influence the continuation of the movie. The interactive film has five different endings mapping out every possible ending, making watching the film like a game. Netflix used artificial intelligence (AI) to create an innovative new way to experience online streaming where views can shape their own story as they go ("Black Mirror: Bandersnatch" Could Signal the future for A. I Directors I Inverse)		
Stage	In market		
Beneficiaries	Subscribers and viewers	• Engaging experience • Personalized experience	

Case Study 3

3D Mapping

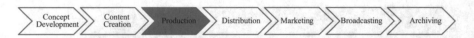

- Countries of implementation: Global
- Company: Green Hippo

Green Hippo develops high-performance video processing software for over 15 years, building quality applications for the entertainment industries. Their core technology is to integrate multimedia products such as projectors and screens through its media servers to create 3D mapping effects and visualizations (About Green Hippo | Green Hippo).

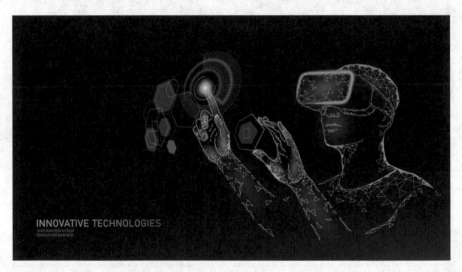

Technology	Augmented reality (AR)/virtual reality (VR)/3D image projection	
Description	Hippotizer offers components enabling intricate mapping projects. Using shape for 3D mapping features, the projection can be aligned, allowing for elaborate LED designs providing a new perspective to real-time video creation with easy-to-set-up interphase	
Stage	In market	
Beneficiaries	Show light designers, facade lighting designers, corporate events organizers	• State of the art 3D effects and visualizations • Design manipulation

Case Study 4 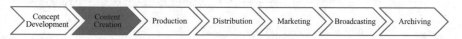 **NIANTIC**

Augmented Reality (AR) in Games

Concept Development | Content Creation | Production | Distribution | Marketing | Broadcasting | Archiving

- Countries of implementation: Global
- Company: Niantic

Niantic is a software development company in the USA and is known for its innovative development and creation of AR mobile games (A Brief History of Niantic Labs, the Makers of Pokémon Go | TechCrunch).

Technology	Augmented reality (AR)/GPS: it is a location-based game
Description	Pokémon Go is an augmented reality mobile game. It uses the GPS in the mobile devices to help locate battles, capture, and train the virtual creators, making the game seem like part of the players' real world. It excited the use of AR and location-based technology to encourage physical activity. The application targeted Pokémon fans to search for Pokémon in the real world (Is Pokémon Go Augmented Reality? | Scientific American)
Stage	In market

| Beneficiaries | Gamers | • Supports physical activity
• Enhanced real-life social exchanges |
| | Employees | • Interactive immersive experience
• Improved training and educational
• Increased potential for remote work and collaboration |

Case Study 5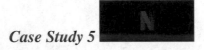

Streaming

- Countries of implementation: Global
- Company: Netflix, Inc.

Netflix, Inc. is an American media services provider and production company, founded in 1997. The company's primary business is its subscription-based streaming service which offers online streaming of a library of films and television programs, including those produced in-house (About Netflix I Netflix Media Center).

| Technology | Platforms/streaming media |
| Description | Online streaming of a library of films and television programs for adults and children |

Stage	In market	
Beneficiaries	Subscribers	• Content variety • Cost-effective • Multiplatform compatibility

Case Study 6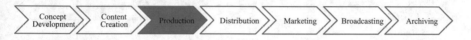

3D Surround Sound

- Countries of implementation: Global
- Company: Dolby Laboratories

Dolby Laboratories, Inc. is an American company specializing in audio noise reduction and audio encoding/compression. Dolby licenses its technologies to consumer electronics manufacturers (Discover Dolby | Dolby).

Technology	Augmented reality (AR)/virtual reality (VR)/3D surround sound
Description	Dolby Atmos is a suite for immersive audio, having both horizontal and vertical sound placement. This technology has been used in recent years in the cinema industry, games, and Netflix. It has provided an opportunity for moviemakers to create a revolutionary sound experience (Dolby Atoms: What is it? How Can You Get it? | WhathiFi)
Stage	In market

Beneficiaries	Consumers	• Immersive sound experience • 3D sound projection • High-quality auditory performance

Case Study 7

Bose Frames

Concept Development	Content Creation	Production	Distribution	Marketing	Broadcasting	Archiving

- Countries of implementation: Global
- Company: Bose Corporation

Bose Corporation is an American manufacturing company that sells audio equipment and produces audio gadgets. The company is known for its wide-range audio systems. Bose is best known for its home audio systems and speakers, noise-canceling headphones, professional audio products, and automobile sound systems (About Us | Bose).

Technology	Augmented reality (AR)/virtual reality (VR)/audio sunglasses	
Description	Bose Frames are wearable glasses that include noise-canceling, miniaturized speakers, and a microphone for personal listening and speech. The wearer can hear the sound coming out of the sunglasses, with minimized sound leak while others hear practically nothing the sound technology directs sound at the wearer and away from others. The device is paired with the smartphone via Bluetooth to play music, receive calls, and add information to what the wearer sees. The device accepts gesture controls for volume control and phone answering. It can interact with virtual assistants such as Siri and Google Assistant. The device also connects to augmented reality mobile applications and includes a rechargeable battery (Boss Frames Review | TechRadar)	
Stage	In market	
Beneficiaries	Consumers	• Multipurpose device • Guided access • Augmented reality application

Case Study 8

Virtual Museums

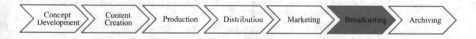

Concept Development — Content Creation — Production — Distribution — Marketing — Broadcasting — Archiving

- Countries of implementation: Global
- Company: Google LLC

Google LLC is an American multinational technology company that specializes in internet-related services and products, which include online advertising technologies, a search engine, cloud computing, software, and hardware.

Technology	Augmented reality (AR)/virtual reality (VR)/virtual museums
Description	Google Arts & Culture is an online platform through which the public can view high-resolution images and videos of artworks and cultural artifacts from partner cultural organizations throughout the world. The digital platform utilizes high-resolution image technology that enables the public to virtually tour partner organization collections and galleries and explores the artworks' physical and contextual information. The platform includes advanced search capabilities and educational tools (When Museums Become Virtual I In exhibit)
Stage	In market
Beneficiaries	Consumers • Increased access to art • Improved visitor experience • Engaging atmosphere

Case Study 9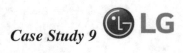

Nano Cell TV

- Countries of implementation: Global
- Country: LG Corporation

LG Corporation is a South Korean multinational conglomerate corporation. It is the fourth largest chaebol in South Korea. LG makes electronics, chemicals, and telecom products and operates in over 80 countries (The History of LG | UKEssays).

Technology	Nanomaterials/8K display resolution	
Description	8K Nano Cell TV offers a real 8K LED experience. It is four times the resolution of 4K, complete with the natural, lifelike color of Nano Color, precise color at wide angles with Nano Accuracy, and deeper black and contrast of precisely balanced lighting. With deep learning, any content resolution can be upscaled to 8K. LG's 8K processor uses artificial intelligence (AI) and deep learning to make it possible, translating the source to over 33 million pixels (What Is 8k and What Are Its Benefits? | The Telegraph)	
Stage	In market	
Beneficiaries	Consumers	• Enhanced viewing • Better imagery

Case Study 10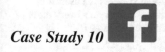

Mixed Reality Oculus Quest

- Countries of implementation: Global
- Company: Facebook Inc.

Facebook is an American online social media and social networking service. Oculus is a division of Facebook Technologies, LLC, a subsidiary of Facebook Inc. Oculus specializes in virtual reality hardware and software products. It was founded in 2012 and acquired by Facebook in 2014 (Facebook to Acquire Oculus | Facebook).

Technology	Augmented reality (AR)/virtual reality (VR)/sensors and computer vision	
Description	Oculus Quest is a revolutionary, wireless, virtual reality headset that could be used alone or plugged into the computer to enable access for games that are not in the headset. Oculus' innovative software delivers a high-quality gaming experience with realistic precision. It is considered an all-in-one gaming headset because it includes visual display, audio, and in-hand controllers, has built-in storage, and does not require a PC. The technology consists of Insight, which is capable of 3D mapping your virtual space for an accurate representation of your surroundings in the real world. This Guardian system prevents the user from hitting an obstacle and has built-in sensors that translate the user's every move into virtual reality (Oculus Quest: Everything you Need to know | AndroidCenter).	
Stage	In market	
Beneficiaries	Gamers	• State of the art • Hands-free • Eliminated need for PC purchase • Cost-effective

Case Study 11 The Washington Post

Artificial Intelligence (AI) in Journalism

- Country of implementation: USA
- Company: The Washington Post

The Washington Post is a significant daily newspaper published in Washington, D.C. The company stands out with its works on aspects of the US government.

Technology	Augmented reality (AR)/virtual reality (VR)/robot reporting program	
Description	The Heliograf is used to produce reports by recognizing a trend in finance and big data. The Washington Post has used this technology to help journalists with reporting. For instance, the Post used Heliograf during the elections to alert journalists when the election results started trending in an unexpected direction. This gave their staff an advantage (The Rise of the Robot Reporter I the New York Times)	
Stage	In market	
Beneficiaries	Reporters, staff	• Time-effective • Increased efficiency • Increased productivity • Improved quality

Case Study 12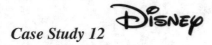

Virtual Theme Parks

- Countries of implementation: Global
- Company: The Walt Disney Company

The Walt Disney Company, more commonly known as Disney, is an American mass media and entertainment business. The company is leading in the American animation industry and lately has diverted to live-action film, production, television, and theme parks (Disney Company | Britannica).

Technology	Augmented reality (AR)/virtual reality (VR)/mobile application
Description	Disney has created an app called Play Disney Parks that enables visitors to play games based on the Disney theme park. It created a unique experience that brings the environment surrounding the player to live. It was developed to entertain guests while waiting to queue to ride Space Mountain at Disneyland etc can design their rocket ships while they wait (Play Disney Parks App Relives Bordon in Line | USA Today)

Stage	In market	
Beneficiaries	Consumers	• Enhanced visitor experience • Entertainment in queue

Case Study 13

Blockchain-Based Live Streaming Platform

Concept Development 〉 Content Creation 〉 Production 〉 Distribution 〉 Marketing 〉 Broadcasting 〉 Archiving

- Countries of implementation: Global
- Company: DLive

DLive TV is one of the leading blockchain-based live streaming platforms. It changed the game for content creators by putting platform ownership in the user's hand using blockchain technology (DLive, a Controversial Blockchain-based Live Streaming Platform | Blockchain News 2020).

Technology	Blockchain/platforms	
Description	DLive firm created a controversial blockchain-based streaming platform monetized by Lino blockchain. It allows users to build their content and promises incentives to users, viewers, and content creators. Users are rewarded for contributing as the platform grows. It revolutionized distribution systems by awarding the platform community for their contribution rather than the corporation (DLive, a Controversial Blockchain-based Live Streaming Platform I Blockchain News 2020)	
Stage	In market	
Beneficiaries	Content creators and contributors	• Low fees • Cost-effective • Enhanced quality • Live to stream • No creator fees • Efficient

Case Study 14

Big Data in Media and Entertainment

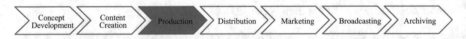

- Countries of implementation: Global
- Company: Qubole

Qubole platform is a secure, open-source, multi-cloud data lake platform that provides services that aid in diminishing time and efforts toward running data pipelines, machine learning, streaming analytics, and data exploration (About Us I Qubole).

Technology	Big data/machine learning and cloud
Description	Qubole provides services to companies that are faced with the pressure of creating new business strategies in the digital world. Production, advertising, and distribution plans rely on understanding and analyzing consumers' behaviors and preferences. As the world shifts from analog to digital media, it gave rise to opportunities for business to leverage their significant data assets to better customer engagement. It predicts what consumers want by collecting data on consumers' most viewed content and devices they use to see them (Big Data in Media and Entertainment I Qubole)
Stage	In market
Beneficiaries	Consumers

Beneficiaries	Consumers	• Improved customer experience • Enhanced customer engagement and loyalty • Constant updates

Case Study 15

Cognitive Movie Trailer

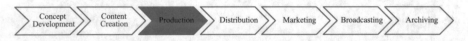

- Countries of implementation: Global
- Company: 20th Century Studios

20th Century Studios is an American film studio that is part of the larger company Walt Disney Studios. It is one of the biggest film studios worldwide.

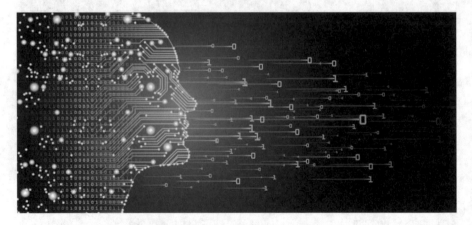

Technology	Artificial intelligence (AI)/machine learning and platform
Description	The company created trailers with the help of machine learning algorithms and artificial intelligence (AI) using IBM's Watson platform. In horror movies, Watson was used recently where the platform survey and understand what creates a sense of horror within the viewers. The platform then selects the topmost workable clips in the films, and the editors use those clips in creating the trailers for them (Machine Learning Already Changing the Entertainment Industry I Future 2016)
Stage	In market
Beneficiaries	Consumers	• Improved quality • Unique, engaging experience
	Employees	• Enhanced workforce expertise and creativity • Effective work management

Case Study 16

Personalized Content

- Countries of implementation: Global
- Company: Netflix

Netflix, Inc. is an American media services provider and production company founded in 1997. The company's primary business is its subscription-based streaming service which offers online streaming of a library of films and television programs, including those produced in-house (About Netflix | Netflix Media Center).

Technology	Artificial intelligence/machine learning and platform	
Description	Netflix created an AI algorithm called Meson that helps the team predict what consumers want to watch. Once the subscriber opens the Netflix platform, TV series, movies, or documentaries that he might want to watch are offered on screen (Machine Learning Already Changing the Entertainment Industry | Future 2016)	
Stage	In market	
Beneficiaries	Consumers	• User friendly • Engaging experience

Case Study 17

The Automation of Creativity

| Concept Development | Content Creation | Production | Distribution | Marketing | Broadcasting | Archiving |

- Countries of implementation: Global
- Company: McCann

McCann is a global American advertising and marketing organization. It is one of the largest industries worldwide and has created iconic adverting campaigns (About Us | McCann).

Technology	Artificial intelligence (AI)/machine learning
Description	McCann Erickson in Japan created an artificial intelligence (AI) creative director AI-CD. Its purpose is to study the buyer's brief then produce creative ideas. It is the first robot designed to provide creative directions for commercials. It should be able to do a creative director's job more effectively (Machine Learning Already Changing the Entertainment Industry | Future 2016)

Stage	In market	
Beneficiaries	Consumers	• Creative, engaging media campaigns
	Employees	• Creative direction • Time-effective • Cost-effective • Enhanced learning experience

Oil and Gas

Abstract The oil and gas industry is one of the world's largest sectors and contributes significantly to the global economy. Oil and gas have become essential sources of energy for industries, homes, and transportation for almost a century. In recent years, we see a decline in the oil and gas industry. Therefore, companies are looking to innovate and develop technologies to increase productivity and reduce costs (The Oil and Gas Industry in the Energy Transitions 2020).

Keywords Self-learning · Autonomous robots · Leak detection · Remote monitoring · Built-in sensors · Payment solutions · Hydrocarbon · Limit pollution · Waste oil · Recycle · Performance improvement · Temperature-sensitive equipment · Operational efficiency · Virtual AI system · Smart manufacturing · Autonomous robots · Tracking platform · 3D printing · AR smart glasses · Geophones · Voice and data connectivity

The oil and gas industry is one of the world's largest sectors and contributes significantly to the global economy. Oil and gas have become essential sources of energy for industries, homes, and transportation for almost a century. In recent years, we see a decline in the oil and gas industry. Therefore, companies are looking to innovate and develop technologies to increase productivity and reduce costs (The Oil and Gas Industry in the Energy Transitions 2020).

There are significant challenges in the oil and gas industry, but thankfully, it comes when this industry is met with technological opportunities. One of the industry's problems is to produce refined industrial products and oil at low prices. Therefore, companies of such sector need to improve environmental utilities and production line on the functioning sites. This will help increase productivity and reduce extraction and refining costs. Another challenge this industry faces is trying to improve performance by extending the life span of mature sites to maintain their supply of crude oil and gas and to seek new sources of raw materials for which extraction, refining, and transportation are more complex and expensive. Companies' target is for their plants to be highly reliable, with an increase in output (Three Major Challenges for the Oil & Gas Industry | Veolia).

© Springer Nature Switzerland AG 2023
M. Anis et al., *Mapping Innovation*,
https://doi.org/10.1007/978-3-030-93627-3_9

Additionally, as the oil and gas industry deflect deep water to meet the world's energy demand, problems such as personnel health, safety, and pollution have been brought up. An unforgettable incident that sparked this attention was the explosion of the Deepwater Horizon offshore petroleum platform that led to the oil being spilled in the ocean (Robotization of Operations in the Petroleum industry | Science Publishing Group, 2019).

Even though nowadays, oil and gas continue to be the dominant source of energy, the developments in renewable energy over the past few years have been the industry's greatest challenge. There is a rise in innovative projects such as electric cars, natural gas vehicles, and a notable cost reduction in solar, wind, and nuclear energy. Nowadays, France uses nuclear power to produce electricity. This threat decreased the reliance on fossil fuels, which are already short in supply, reduced the need for dangerous gas, and eliminated the need to face adversity against the harsh working conditions of this industry (Renewable Energy Resources in the Egyptian Oil and Gas Industry: Outlooks and Challenges | Research Gate, 2018).

Finally, the industry encountered a price shock in recent years due to the low demand for oil and gas. The current context of the COVID-19 crisis has also led to a collapse in oil prices and a decline in its demand. It became essential to put efforts into adopting new ways to help businesses survive and become more resilient. There is a considerable decline in oil consumption since transportation is at a virtual standstill, as many governments are asking their people to stay home (COVID-19: What's Next for Oil and Gas | Publicis Sapient).

The opportunities that lie ahead in the oil and gas industry are tremendous. There is a clear indication that the world will remain dependent on those raw materials for several decades. This industry has the opportunity to innovate and develop technologies. New technologies will be a significant factor in finding and creating new sources of hydrocarbons, such as oil and gas, in the most demanding environment. For example, shale gas is being developed in North America to generate electricity. Hydraulic fracturing technology is used to safely unlock gas resources trapped in small rock spaces such as shale gas. Additionally, due to the current pandemic, people are investing in emerging technologies that aim to improve productivity, reduce risk, and cut costs in the long run. Many technological innovations aim to overcome the adversity of the oil and gas industry (Why Is Shale Gas Important?).

Value Chain

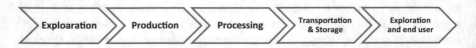

The value chain of this industry is comprised of five phases: exploration, production, processing, transportation and storage, and end user.

In the exploration phase, the oil and gas companies evaluate different locations for drilling, field development, and extraction operations. Then during the production phase, they start the extraction process itself. The resources obtained are then processed and transformed during the processing phase so they can be sent to be used by the consumer. The last phases are transportation and storage and finally end user phase.

Case Studies

Case Study 1 ExxonMobil

Artificial Intelligence (AI) and Automation in Self-Learning Submersible Robots

- Country of implementation: USA
- Company: ExxonMobil LNG

ExxonMobil LNG is a part of the ExxonMobil Corporation and is the largest international oil and gas company. They aim to use innovative technology to help satisfy the world's natural gas and power demands. It has one of the largest refiners and marketers in the oil and gas industry.

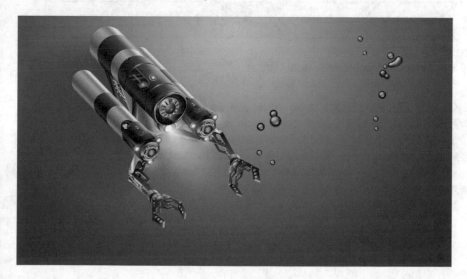

Technology	Automation/computer vision with 3D visual mapping cameras
Description	ExxonMobil is working on self-learning submersible robots to investigate and enhance natural seep detection abilities at sea. Natural seeps occur when oil escapes from rock found on the ocean floor. With many shallow-water fields exhausted, oil and gas are now being extracted from deep waters. Autonomous robots for ocean applications are a lot like self-driving cars. The robot's leading role is to do the day and night inspections, which are currently fulfilled by humans, detect any unusual activities such as malfunction or leak, and intervene in case of an emergency or whenever necessary. Seafloor details, such as photos, measurements, and locations, are collected and analyzed. Researchers use autonomous underground vehicles with 3D visual mapping cameras to gather such information, find biological hotspots, and target them for sampling and observation. Unsupervised algorithms may be used to form deployment plans (GEOEXPRO, 2019)
Stage	Research phase

Beneficiaries	Researchers and employees	• Increased efficiency • Increased productivity • Increased safety • Reduced costs • Increased efficacy • (AI for Ocean Exploration: Is an AI Colony Possible in the Deep-Sea I electronics, 2019)

Case Study 2

Big Data and Data Analytics in Oil and Gas

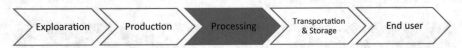

- Countries of implementation: Global
- Company: General Electric Digital

General Electric Digital is an American multinational subdivision of General Electric and one of the early endorsers of big data and data analytics. The company's primary goal is to provide innovative solutions to challenges worldwide industrial companies face with software and the internet of things (IoT) (industrial internet of things). It explores how assets in this industry are run, constructed, and maintained and used by machine learning, the internet of things (IoT), and big data to convert innovative insights into profitable and productive company outcomes.

Technology	Big data analytics/internet of things (IoT) and machine learning
Description	GE Digital developed a digital platform called Predix which is used to create software that represents the company's physical assets. This software is called the Digital Twin. The industry produces a vast amount of data that is difficult to analyze and manage. The platform's main purpose is to provide industrial companies with fast, secure, and effective internet of things (IoT) applications to turn their asset data into pragmatic insights. The platform's machine learning cognition processes a large amount of data collected from sensors, compares them to ideal parameters found in the database, and looks for any inconsistency between the collected and ideal states. If any inconsistency is found, the platform sends a warning to the technicians to fix the issue. With this technology companies can better comprehend, forecast, and optimize their assets production. Many leading industrial companies, such as Schneider and Maersk, are using this innovative solution to enhance efficiency and improve the operation of their assets (GEOEXPRO, 2019)
Stage	In market
Beneficiaries	Customers

Beneficiaries	Customers	• Security of data • Legal compliance
	Employees	• Improved industry safety • Enhanced asset operations • Cost-effective

Case Study 3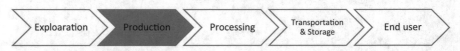

Internet of Things (IoT) in Oil and Gas Industry

| Exploaration | Production | Processing | Transportation & Storage | End user |

- Countries of implementation: Global
- Company: Intelligent Sensing Anywhere

It is an internet of things (IoT) company for the oil and gas market that has been in the field for 28 years implementing remote monitoring solutions.

Technology	Internet of things/servers, networks, and cloud	
Description	The company focused on recording and transmitting the readings of the wireless sensors equipped on fuel and gas tanks and connected to the internet to enable remote monitoring of tank exploitation for companies working in the oil and gas industry	
Stage	In market	
Beneficiaries	Clients	• Detailed tank exploitation • Ease of access to the instrument reading
	Employees	• Ease of access and remote monitoring • Reduced working hours and effort

Case Study 4 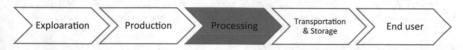 sparkcognition™

Predictive Maintenance of Machinery Using AI and Internet of Things (IoT)

Exploaration	Production	Processing	Transportation & Storage	End user

- Countries of implementation: Global
- Company: SparkCognition

This company offers artificial intelligence (AI) applications to solve certain challenges for other institutions. It is based in Texas, USA, and was founded in 2013.

Technology	Artificial intelligence/internet of things (IoT)	
Description	SparkCognition created a set of integrated applications that can anticipate malfunctions in the performance of the machines. The first application is called DeepNLP, which collects the data from various sources to prepare for analytics. The second application is SparkPredict, which is the machine learning part responsible for analyzing the data using algorithms to discover failures. Then, Darwin, the third application, builds the model of prediction by training it on the gathered data. Finally, DeepArmor is an application that detects viruses and malware that might enter the system	
Stage	In market	
Beneficiaries	Clients	• Identification of necessary maintenance • Lower downtown • Increased reliability

Case Study 5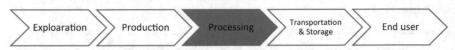

ANYmal Offshore Robot for Challenging Terrain

| Exploaration | Production | Processing | Transportation & Storage | End user |

- Countries of implementation: Global
- Company: ANYbotics

It is a company that develops robots for industrial applications. It is based in Switzerland, and it was founded in 2016.

Technology	Automation/artificial intelligence (AI)
Description	ANYmal is a robot manufactured by ANYbotics. It can inspect offshore sites using built-in sensors, cameras, and microphones. It is also capable of generating a 3D map of its surroundings. It can be remotely controlled from a control unit to receive the real-time readings from the robot
Stage	In market
Beneficiaries	Clients

Beneficiaries	Clients	• Remote controlled
		• Efficient tasks completion
		• Employee safety

Case Study 6 IBM

IBM Watson Using Big Data Analytics to Solve Problems in Oil and Gas

| Exploaration | Production | Processing | Transportation & Storage | End user |

- Countries of implementation: Global
- Company: IBM

It is an American multinational company that specializes in technology. It is a leading company in the field of artificial intelligence (AI) and cloud platforms. It is was founded in 1911, in New York City, USA.

Technology	Big data analytics/artificial intelligence (AI), platforms	
Description	IBM created a platform called IBM Watson that acts as a robotic advisor for operation engineers who need to make decisions when a certain problem is encountered in a site. For example, when water pressure increases in a well, Watson sends relevant information from a similar event that happened before and how did the decision before affecting the utilities. This can be extremely helpful in better decision-making	
Stage	In market	
Beneficiaries	Operation managers	• Improved decision-making

Case Study 7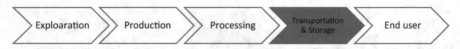

Getting to Value with Blockchain in Oil and Gas

| Exploaration | Production | Processing | Transportation & Storage | End user |

- Countries of implementation: Global
- Company: Accenture

Accenture is an Irish multinational company specialized in technological services, cloud services, and digital marketing since 2009. It has been serving clients in more than 200 cities, in 120 countries.

Technology	Blockchain/artificial intelligence(AI), platforms
Description	Accenture's track and pay solution allows the recording of thousands of transactions that occurred by truck, rail, plane, and ship. The use of this blockchain solution decreases the discrepancy of detailed analysis for costs by providing a single source for access through a distributed database
Stage	In market
Beneficiaries	Clients

Beneficiaries	Clients	• Cost savings • Increased visibility and transparency • Accuracy of freight invoices

Case Study 8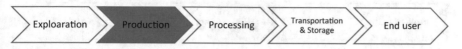

Making the Oil and Gas Industry Greener

| Exploaration | Production | Processing | Transportation & Storage | End user |

- Country of implementation: USA
- Company: Imaginea

It is a Silicon Valley-based company keen on providing product development and engineering services to solve technology problems. It was founded in 2009, and today it has more than 1000 highly skilled engineers spread across the globe.

Technology	Green technologies/ecosystems	
Description	Imaginea utilizes a clean hydrocarbon ecosystem to provide production of energy that minimizes the usage of freshwater and without toxic emissions that pollute the air.	
Stage	In market	
Beneficiaries	Society	• Clean living environment

Case Study 9

Recycling in the Oil and Gas Industry

| Exploaration | Production | Processing | Transportation & Storage | End user |

- Countries of implementation: Global
- Company: EnerPure

HD-Petroleum became EnerPure to reflect the reality of the company being a clean-tech energy company, a company that recycles waste petroleum into a clean source of fuel and energy while reducing greenhouse gas emissions.

Technology	Green technologies/recycling	
Description	The company has created waste oil refinery units to convert them to diesel fuel. The implementation cost of the technology is relatively low, and the process of recycling decreases the gas emissions	
Stage	In market	
Beneficiaries	Society	• Cleaner habitat • Ecofriendly
	Oil and gas companies	• Better oil disposable methods • Increased in revenue through recycling

Case Study 10 *Ingersoll Rand*

Internet of Things (IoT) System for Failure Alert in Machines

| Exploaration | Production | Processing | Transportation & Storage | End user |

- Countries of implementation: Global
- Company: Ingersoll Rand

Ingersoll Rand is a global company that provides innovative solutions in air, fluid, energy, and medical technologies. It has merged with Gardner Denver in early 2020.

Technology	Internet of things (IoT)/platforms, artificial intelligence (AI)	
Description	Sensors connected to valves, pumps, and other machines gather information. This information is then transmitted to the cloud via an internet connection. The data gets stored and analyzed by machine learning algorithms. The algorithms have been already trained on the normal functionality of the machines so that they can know when a fault happens and alert the maintenance team	
Stage	In market	
Beneficiaries	Oil and gas companies	• Improved equipment performance • Employee safety

Case Study 11 PERCEPTO

Inspection Drones for the Oil and Gas Industry

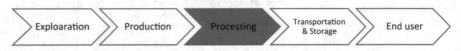

| Exploaration | Production | Processing | Transportation & Storage | End user |

- Countries of implementation: Global
- Company: Percepto

It is a company that builds autonomous drones for industrial sites, and it has deployed systems across the Americas, Europe, and Asia to solve critical operational challenges.

Technology	Artificial intelligence/cloud, computer vision, platforms
Description	Percepto's solution is a drone that leaves its base to collect data from sites, send them to the cloud, and return to the base for charging. The key applications are scheduling 24/7 routine patrols of site perimeters for security, monitoring temperature-sensitive equipment with thermal mapping technology, and detecting leakage
Stage	In market
Beneficiaries	Oil and gas companies

• Improved equipment performance
• Improved employee safety

Case Study 12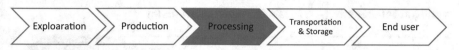

On-the-Edge Maintenance Device

| Exploaration | Production | Processing | Transportation & Storage | End user |

- Countries of implementation: Global
- Company: Foghorn

It is a company that offers software called Lightning Edge Intelligence that helps oil and gas businesses carry out maintenance with an automated system for detection and repair using machine learning technology.

Technology	Artificial intelligence (AI)/automation	
Description	The majority of the AI solutions send collected data from the site to a cloud where the analysis happens on a remote server. However, some sites lack a good internet connection for the transmission of this large amount of data that include images and videos. Foghorn offers a solution that analyzes the system placed in the site with even a footprint of less than 256 MB. The system is capable of detecting operational efficiency for pumps and pipelines. If a potential malfunction gets detected, the system can stop the pump to prevent damage or shutdown valves	
Stage	In market	
Beneficiaries	Clients	• Balanced reliability • Optimized KPIs • Reduced manual data entry • Lowered maintenance costs

Case Study 13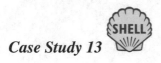

Artificial Intelligence (AI) in the Oil and Gas Industry

| Exploaration | Production | Processing | Transportation & Storage | End user |

- Countries of implementation: UK and USA
- Company: Royal Dutch Shell

Royal Dutch Shell is international cooperation specializing in the exploration, production, and distribution of oil and natural gas. Its goal is to produce innovative energy solutions economically and environmentally to meet the growing population's demand for energy (Who We Are I Shell Global).

Technology	Artificial intelligence (AI)/online chat platform
Description	Shell is the first company in the lubricant sector to launch a virtual AI system for customers. It provides a solution for customers who spend a lot of time and effort maneuvering through an extensive database searching for products in the lubricant sector. The system consists of two avatars, Emma and Ethan, to help customers search for and discover products using spoken language. The system is accessible using an online chatting platform on the company's website. It provides information about where a specific product could be found, specifications and information about the product, and the amount available (Artificial Intelligence (AI)in Oil and Gas - Comparing the Application of Oil Giants I Emerj)

Stage	In market	
Beneficiaries	Customers	• Customers reach their product faster, easier, and more effectively • Round-the-clock efficiency
	Employees	• Improved time management and efficiency • (To Hire or Not to Higher: The Advantages and Disadvantages of Virtual Assistance | Square Fish)

Case Study 14

How Oil Giants Are Using Artificial Intelligence (AI)

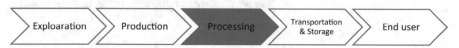

- Country of implementation: China
- Company: Sinopec

Sinopec is a Chinese chemical and petroleum company and one of the largest oil companies. It is one of the two cooperatives that control the Chinese oil and gas business. It produces resources and develops petroleum products. The company aims to meet China's energy demand and now is directed to the domestic development of shale gas (China Petroleum & Chemical Corp | Asian Review).

Technology	Artificial intelligence (AI)/platform and internet of things (IoT)	
Description	Sinopec is trying to implement smart manufacturing to enhance efficiency and make the company more robust. The platform used AI to provide a focal technique of managing data and its fusion in different applications used to manage the plant's performance. The AI technology would construct modules and regulations that would notify how data is comprehended and present opportunities for actionable insights that enhance the factory's production (Artificial Intelligence (AI) in Oil and Gas – Comparing the Application of Oil Giants	Emerj)	
Stage	In market	
Beneficiaries	Customers	• Faster and more informed decision-making • Better quality and efficiency of the end product
	Employees	• Easier and more efficient, real-time task monitoring • Task analysis • Recommended actionable solutions • Improved safety • labor productivity • (Sinopec's Future Development | Fei Meng Automation)

Case Study 15

Artificial Intelligence (AI) in Hydrocarbon Exploration

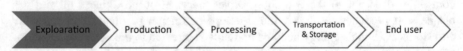

- Countries of implementation: Global
- Company: Total

Total is one of the most significant oils, natural gas, and petrochemical corporations worldwide. It is a French-based company that covers the whole oil and gas chain. It specializes in the exploration and manufacturing of crude oil and natural gas as well as advancements in gas electricity and operation activities in coal mining. Additionally, it produces chemical products and the advertising, selling, and refining of oil and gas products (History of Total S.A | Reference for Business).

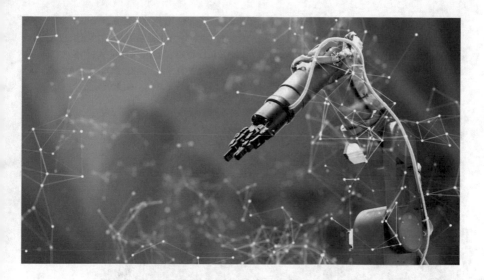

Technology	Artificial intelligence (AI)/automation and robotics
Description	Total launched a global competition for teams to create prototypes of autonomous robots to present well-rounded reports at exploration sites. AI's key role in this robot is to provide a report with real-time data and effectiveness analysis in different exploration locations. ARGOS has the potential to fulfill day and night inspections that are currently done by humans. Additionally, it detects any out-of-the-ordinary activities and fixes any major issue (Artificial Intelligence (AI) in Oil and Gas - Comparing the Application of Oil Giants | Emerj)
Stage	Prototyping phase
Beneficiaries	Employees

Beneficiaries	Employees	• Safer work environments • Labor-saving • Improved efficiency and productivity

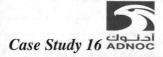

Case Study 16 ADNOC

Blockchain in Oil and Gas Operations

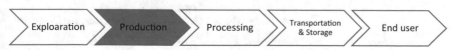

Exploaration   Production   Processing   Transportation & Storage   End user

- Countries of implementation: Global
- Company: ADNOC

ADNOC is one of the largest energy producers in the world. The company covers all the functions in the oil and gas value chain such as exploration, refining, advertising, distribution, and production one petroleum chemicals. They play a vital role in Abu Dhabi's economy where the company is one of the major contributors to the city's economic prosperity and diversification (Who We Are I ADNOC).

Technology	Blockchain/automation and tracking systems	
Description	ADNOC, a UAE-based oil giant, plans to adopt blockchain technology in crude oil and natural gas production to improve efficiency and performance. Blockchain technology provides a safe and secure platform for tracking, verifying, marketing, and streamlining daily transactions. It also reduces the company's operation costs by reducing time consumed during sales and improves data reliability during production. It can keep a note of how much oil is transmitted from the production well to the refinery and how much gas is being exported abroad or traded between firms and the costs of doing such functions (ADNOC and IBM Develop Blockchain or Oil and Gas Operations I Offshore Technologies)	
Stage	Developing phase	
Beneficiaries	Customers	• Faster, safer, and more convenient fuel sales • Enhanced stakeholders and customers clarity and transparency
	Employees	• Increased safety • Improved maintenance • (Oil and Gas Industry - Blockchain, the Disruptive Force of the 21st Century I Infosys)

Case Study 17 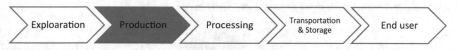 GE Digital

3D Printing in the Oil and Gas Industry

| Exploaration | Production | Processing | Transportation & Storage | End user |

- Countries of implementation: Global
- Company: General Electric Digital

General Electric Digital is an American multinational subdivision of General Electric and one of the early endorsers of big data and data analytics. The company's primary goal is to provide innovative solutions to challenges worldwide industrial companies face with software and the internet of things (IoT). It explores how assets in this industry are run, constructed, and maintained and used by machine learning, the internet of things (IoT), and big data to convert innovative insights into profitable and productive company outcomes.

Technology	3D printing/platforms	
Description	The company's 3D printing services allow oil and gas companies to accelerate product development. Therefore, they will be able to test what they design rapidly to reduce the time taken for full production. They will also be able to create complex geometries to meet maximum performance measures. Moreover, they can manufacture spare parts easily and efficiently	
Stage	In market	
Beneficiaries	Customers	• Maximum production potential • Rapid prototyping of machinery, like turbines

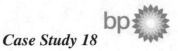

Case Study 18

Augmented Reality in the Oil and Gas Industry

| Exploaration | Production | **Processing** | Transportation & Storage | End user |

- Country of implementation: UK
- Company: British Petroleum

BP is a multinational oil and gas company headquartered in London, UK. It is one of the world's seven oil and gas "supermajors," whose performance in 2012 made it the world's sixth largest oil and gas company.

Technology	Augmented reality (AR)/internet of things (IoT)	
Description	BP involved the usage of AR smart glasses that can superimpose digital images for technicians to help them fix the problems they face in sites. The glasses also transmit real-time data to the control unit where engineers can reply with helpful data for guidance. The smart glasses that BP uses are designed by a company called "Fieldbit." The company also states that the technology can assist in error detection and correction and also in preventing oil spills	
Stage	In market	
Beneficiaries	Customers	• Reduced maintenance costs
	Employees	• Technical support

Case Study 19 CNPC

Digital Seismic Mapping for Petroleum

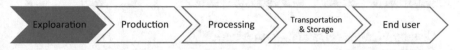

Exploaration > Production > Processing > Transportation & Storage > End user

- Country of implementation: China
- Company: China National Petroleum

Established in 1988 and the predecessor of the Ministry of Petroleum Industry of the People's Republic of China, the China National Petroleum Corporation is a state-owned oil and gas company with governmental administrative functions. In 2018, the company's net income reached $5.4 billion, and it produced 1.9 million barrels per day.

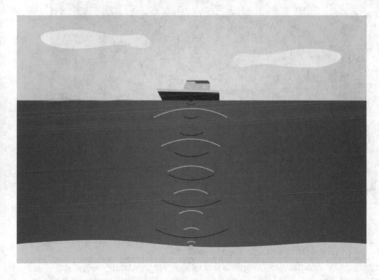

Technology	Artificial intelligence/green technologies
Description	The company uses ultrasensitive sound-emitting devices, called geophones, to help seismologists bounce sound waves off underground rock formations to uncover hydrocarbon reservoirs. The resulting echoes are recorded and converted into three-dimensional maps that are then analyzed by supercomputers that help cut down on the time and money costs of exploration
Stage	In market
Beneficiaries	End user · Fuel availability

Case Study 20 X2nSat

Satellite Communication for Oil and Gas Companies

| Exploaration | Production | Processing | Transportation & Storage | End user |

- Country of implementation: North America
- Company: X2nSat

X2nSat started 20 years ago with consulting for companies that need communications solutions and custom networks. Over the years, it has become a full-service satellite network operator.

Technology	Internet of things (IoT)/satellite communication
Description	The company uses the latest satellite technology for providing oil and gas companies with voice and data connectivity throughout the supply chain of the industry. Additionally, it helps in the transmission of sensor data from the offshore sites to the control units. The company's solutions are scalable and reliable for harsh environmental conditions
Stage	In market
Beneficiaries	Clients	• Reliable communication with engineers and technicians on site

Printing and Packaging

Abstract The printing and packaging industry is the process for delivering products from the point of production to the end of utilization. It has several purposes, including protecting the product until it reaches its destination, maintaining the product's value, and eliminating product waste. The industry achieved rapid growth and is expected to expand more and more in the coming years. These packages are usually designed to appeal to and captivate the manufacturer of the product or consumer through printing and branding. In other words, packaging is the process of protecting a product using different packaging formats, such as boxes, containers, bags, and many others. The packaging is responsible for more than a quarter of plastic worldwide; it increases the shelf life of these products and makes food shopping more suitable for billions of people.

Keywords Sustainable quality · Biometrics · Eye tracking · Information capturing · Recycling · Digital data · End-to-end supply chain · PET bottles · Automated machine · Product journey tracking · Monitor machines · Real-time data · Recycled packaging · Increased shelf-life · Reduced human error · Real-time detection · Optimized packaging · Robotic arm · Reusable material · Biodegradable · Environmentally friendly · VR glasses

The printing and packaging industry is the process for delivering products from the point of production to the end of utilization. It has several purposes, including protecting the product until it reaches its destination, maintaining the product's value, and eliminating product waste. The industry achieved rapid growth and is expected to expand more and more in the coming years. These packages are usually designed to appeal to and captivate the manufacturer of the product or consumer through printing and branding. In other words, packaging is the process of protecting a product using different packaging formats, such as boxes, containers, bags, and many others. The packaging is responsible for more than a quarter of plastic worldwide; it increases the shelf life of these products and makes food shopping more suitable for billions of people.

One of the opportunities of this industry is information sharing, such as adding health and safety information on the package or supporting brand loyalty. Moreover, people working in this industry are trying to reduce packaging weight and use reusable packaging to reduce packaging expenditure and provide less waste generation as denser packaging means more carbon emission and more fuel consumption. Another opportunity in this industry is biodegradable packing. It is on the rise; it offers the same functions as standard packaging while having sustainable qualities while reducing pollution and waste production.

On the other hand, there are some challenges in the printing and packaging industry, including rapid changes in technologies where some workers and businesses find it hard to follow, stay up to date, and deal with the rise of the cost of raw materials. In these sustainability challenges, the company might have commitment difficulties in using recycled materials while keeping high-quality standards (McCarthy 2018).

Moreover, the COVID-19 pandemic has created a massive obstruction in the printing and packaging industry. The coronavirus has led to a massive decline in demand for certain types of packaging while at the same time increasing growth for others, such as packing for e-commerce shipments; this decline has dramatically affected many companies. Another major challenge is that companies are in danger of having a shortage of resources or stocks to create packages due to workers in factories being self-isolated and quarantined (Agrawal 2020).

COVID-19 has also raised opportunities for some companies and businesses, such as sustainable quality. Everyone working in companies that use standard plastic is starting to voice their concerns about using plastic packaging in the environment while still bearing in mind that the investment will be very high to carry out the research needed. Moreover, demand for the packaging of groceries, medicine, and e-commerce transportation is increasing rapidly as people are in lockdown (Agrawal 2020).

Value Chain

This industry's value chain comprises five phases: product development and specifications, production, printing and branding, logistics and distribution, point of sale and usage, and recycling.

The first phase of the chain incorporates multiple research and development activities that develop the products to certain specifications on the printing and packaging of the materials. This is followed by the materials being physically produced (production, printing, and branding). Once ready, the next phase takes into account the logistics of how they will get to where they need to be (logistics and distribution) to be sold and used by the consumer (point of sale and usage). Like other various industry chains, it ends with the recycling phase where printed materials are recycled to be reused as and when required.

Case Studies

Case Study 1 ethimedix

Smart Bottle

- Countries of implementation: Europe
- Company: Ethimedix SA

Ethimedix SA is a privately owned company that was founded in 2010. It aims to provide and design medical devices that enhance drug delivery while improving compliance and at the same time preventing misuse of the medicine (Ethimedix SA).

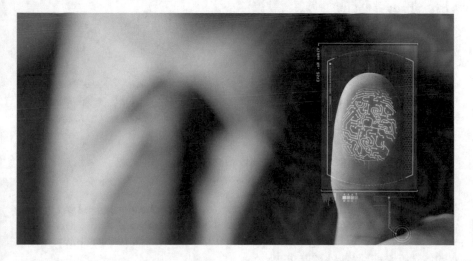

| Technology | Biometrics/fingerprinting technology |

Description	Smart Bottle is designed using biometrics technology where the patients can receive their liquid medication as well as their oral medication. It requires fingerprinting authorization to ensure that only the patients are receiving their prescribed doses. It holds up to one liter of the prescribed medication in an internal cartridge. It can be used both at home and at the hospital or clinic. The prescriber or the physician will be able to track and trace the patient's treatment and the patient will be protected against drug overdosing as using this technology he/she will only be able to take the programmed dose. Moreover, in case of anyone tampering with the medicine or trying to steal it, the system neutralizes it making it unavailable. This means that the medicine will no longer have any effect and cannot be used (Ethimedix)	
Stage	In market	
Beneficiaries	Patients	• Comfortable • Home treatment • Prevention of accidental drug overdoses • Decreased chance of missed dosage • Monitored patient treatment • Decreased risk of product tampering
	Health-care providers	• Easier • Modification of doses • Correct dose administration • Decreased potential for theft and misuse • Medication traced to ensure compliance • Cost-effective

Case Study 2  UNRAVEL

Eye Tracking

- Country of implementation: The Netherlands
- Company: Unravel Neuromarketing Research

Unravel Neuromarketing Research was founded in 2012 headquartered in Utrecht, Netherlands; it aims to predict and increase the effectiveness of a company's marketing as this research program can measure the customer's responsibility to any products by connecting to our brain.

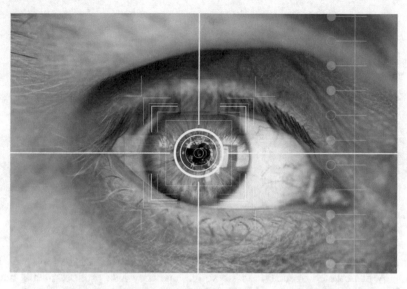

Technology	Biometrics/eye tracking technology, sensors, and electroencephalogram (EEG)
Description	Unravel Neuromarketing Research designed a device that uses eye tracking technology with sensors embedded in it. Researchers will be able to tell by the data received if the customer found a certain packaging or product exciting or not. It will be able to measure or collect data on where the customers looked, how long did they for, and in what order. Moreover, using an electroencephalogram (EEG), which is a monitoring method that allows users to observe the activity of the brain, researchers will be able to tell which emotions or feelings are provoked when looking at a particular package. 4 metrics deliver to the researchers the most information about the person's emotional experience which are desire, engagement, workload, and distraction (Unravel Neuromarketing Research)
Stage	In market
Beneficiaries	Packaging supplier or brand owner	• Feedback of effectiveness of packaging • Feedback on design creativity • Identifiable strengths and weaknesses of packaging

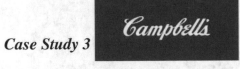

Case Study 3

Measuring and Monitoring of Biological Reactions

- Country of implementation: North America
- Company: Campbell

Campbell is a soup company that was founded by Joseph A. Campbell in 1869. The company has created over 300 classic soups, and they also produce meals, beverages, fresh foods, and snacks with annual sales of over $8 billion (Campbells).

| Technology | Biometrics/sensors and eye tracking technology |

Description	Campbell announced the change of their packaging label in 2010 and the changes that were made were determined by doing a 2-year neuromarketing research by showing to customers pictures of logos, bowls, etc. They were able to monitor and analyze the customer's skin moisture, heart rate, and other biological reactions to know which is the most effective packaging. Moreover, by using eye tracking technology, they were also able to monitor the pupil dilation to track the negative, whether it was a bored or anxious reaction, and positive responses to the package. By using this technology, Campbell found that by using steam in the packaging, customers were more emotionally engaged as the soup looked warm and they also found out that removing the spoon was a better choice as having a spoon in the packaging felt unnecessary to most customers. Moreover, Campbell discovered that in the old label the logo was at the top which many customers found that it drew too much attention which they felt was a negative thing (NeuroRelay 2012)	
Stage	In market	
Beneficiaries	Customers	• More emotionally involved with the packaging • More exciting packaging
	Campbell's designing sector	• Feedback on customer response • More confident about the packaging • Improve the customer's emotional response to the packaging

Case Study 4

Biometric Authentication

- Country of implementation: USA
- Company: Cortegra Group Inc.

Cortegra is a branch of Menasha Corporation which is one of the oldest companies in the country; it was founded in 1849. Cortegra aims to provide and design packaging and labeling services and brand authentication technologies for other industries, such as pharmaceuticals and health care (Turchette).

Technology	Biometrics/proprietary software, camera, digital signature	
Description	Cortegra is a camera-based technology that captures information that is embedded in the packaging (the unique natural microstructures) of the product and then it has proprietary software that transforms it into a digital signature. The digital signature that was obtained is based on the features of the structure; it is a very unique signature, more than a fingerprint with only 1 in 1027 has an identical signature and cannot be duplicated very easily. This technology is used also in high-security markets (Turchette)	
Stage	In market	
Beneficiaries	Pharmaceutical and distribution companies	• The high-security solution that cannot be replicated • Item level authentication • Minimal to no changes or additions to the product and/or package • Applicable to all types of packaging substrates • Merges easily with existing technologies • Real-time registering with no impact on production lines • Nondestructive readings • Utilizes a convenient handheld reader/verifier which allows for easy field authentication • Protecting against fake products • Ensures product security and brand protection

Case Study 5 ⭕ CIRCULARISE

Circularise Plastic

- Country of implementation: The Netherlands
- Company: Circularise

Circularise is a biotechnology company that was founded by Mesbah Sabur in 2016. It aims to enhance the circular economy and eliminate the communication barrier between brands, suppliers, and manufacturers; it allows secure and open communication instead of using blockchain technology.

Technology	Blockchain/digital technology
Description	This technology or communication protocol is available to any business that manufactures, processes use, or recycles plastic. It aims to provide transparent circularity in plastics in a production environment. This technology allows manufacturers to create a digital version of their produced materials using trusted audit documents. This helps manufacturers prove to other businesses and consumers the sustainability of their supply chain which ensures reliability and trust. This technology allows businesses and companies to be more transparent but not to share private information; it ensures a fast transition platform to create a more circular economy for plastics (Poole 2020)
Stage	Testing

Beneficiaries	Consumers	• Ability to track products at any stage of their lifecycle • Informed of the easiest return method
	Recyclers	• Recyclers can turn this waste into high-quality feedstock for new resins, thereby keeping the value of the material as high as possible • Information from previous loops can be linked to new cycles • More efficient decisions on reusing and recycling
	Material producers	• Increased material value • Increased product trust
	OEMs and brand owners	• Achieve sustainability targets • Strengthened brand position • Environmentally friendly

Case Study 6

Product Packaging to Track and Trace Products

- Countries of implementation: Global
- Company: Fasicuchain

Fasicuchain is a blockchain company founded in 2017, and it enhances product tracking and tracing. By using blockchain technology, businesses can benefit from three key features, including authenticity by confirming attributes of a package through supply chains, visibility by allowing businesses to be more transparent and traceable, and finally interactivity by making the retailer, brand owner, and consumers interact with one another (Fasicuchain).

Technology	Blockchain/digital technology and F-code
Description	This company enhances product packaging by connecting the digital data of the package with any physical product. Each package has an F-label printed on it and these F-labels are connected to an F-code which is a cryptographic unique code stored in the blockchain distributed ledger which makes all consumers trust the product. Brands can obtain their private blockchain which will offer them an end-to-end supply chain, from packaging to manufacturing to transportation and finally to the consumer. Brands also have the advantage of ensuring brand loyalty by giving consumers F-coin which is a reward currency. Fasicuchain already has over 500+ packaging and label suppliers (Fasicuchain)
Stage	In market
Beneficiaries	Brand owners and retailers

Beneficiaries	
Brand owners and retailers	• Increase brand loyalty • End-to-end supply chain • Increased consumer trust • Traceable product

Case Study 7 BariQ

PET Plastic Recycling

Product Development & Specs → Production, Printing & Branding → Logistics & Distribution → POS & Usage → Recycling

- Countries and continents of implementation: Egypt, the Middle East, Europe, USA, and Australia
- Company: BariQ

BariQ is a company responsible for the manufacturing of recycled polyethylene terephthalate (RPET) pellets in Egypt. It has the support and technologies of several world-class providers. It was founded in 2010; it aims to bring back meaning to the term green environment and to provide growth opportunities for other businesses (BariQ).

Technology	Green technologies/plastics recycling	
Description	BariQ uses green technologies from other manufacturers in Europe, including Germany, Italy, Norway, and Austria. The company receives PET bottles that are usually used for water, carbonated soda drinks, or oil. The bottles go through processes in different production lines where the first product line prewashes and sorts both automatically and manually to remove all the unwanted materials from them. Then they are crushed into grinders which are then transformed into flakes that are washed with water and chemicals. The second production line uses the produced flakes to turn them into pellets using the extruder then crystallize them. After that, the process of decontamination starts where these pellets will be transformed into food-grade.raisin to make it possible to use them as an input to food contact products, such as bottles, trays, etc. Every year BariQ recycles around 25,000 MT of waste (which are equivalent to 1.6 billion PET bottles) into 15,000 MT of food-grade pellets (BariQ)	
Stage	In market	
Beneficiaries	Brand owners and other businesses	• Meeting sustainability goalsenvironmentally friendly
	Environment	• Environmentally friendly

Case Study 8  TOMRA

Reverse Vending Machines

Product Development & Specs	Production, Printing & Branding	Logistics & Distribution	POS & Usage	Recycling

- Countries of implementation: Global
- Company: TOMRA

TOMRA is an innovation company that was founded by Petter and Tore Planke in 1972; it aims to manufacture or produce solutions for resource optimization to enable a circular economy and minimize waste in the food, mining industries, and recycling. TOMRA collects more than 40 billion empty bottles and provides customers with a productive way of collecting, sorting, and processing these bottles.

Technology	Green technology/recycling
Description	The reverse vending machines allow people to return empty bottles or containers after using them for recycling. The machine then gives the people a refund amount. Hence, this is why it is called a reverse vending machine. These machines are an automated way to collect the return of used drinking bottles. Every year the reverse vending machines in TOMRA receive more than 35 billion used bottles or containers. These vending machines are a very important part of deposit systems as 70% to 100% of all drinking containers are returned for recycling (TOMRA)

Stage	In market		
Beneficiaries	End users		• Convenient recycling locations • Fast and clean machines • Choice of payment methods (some regions offering donations to charities, digital payout via electronic funds-transfer or retail vouchers toward their grocery bill) • Improved carbon footprint
	Environment		• Separate beverage containers by material type (PET, glass, aluminum) • Clean loop recycling • Reduce the need for raw materials in making more containers • Decreased need for landfills • Keeps litter and waste out of groundwater, oceans, and streets
	Staff		• Recycling collection cleaner and streamlined • Reduced time for manually handling and counting containers • More effective time management • Recycling stations take up less physical space on site • Increased storage capacity with compaction • Reduced transport needs for containers • Easier for operators to handle, clean, and maintain

Case Study 9 wikitude

Augmented Reality Product Packing

- Countries of implementation: Global
- Company: Wikitude

Wikitude GmbH is a company that specializes in mobile augmented reality (AR) technologies for tablets and smartphones; it has become the leading company in AR with about 100,000 registered developer users. It was founded in 2008, and in 2012 it announced the launching of Wikitude SDK which is the company's core product. Wikitude SDK uses image recognition, tracking, and geolocation technologies (Wikitude Media).

Technology	Green technology/augmented reality	
Description	Customers can easily download the Herbal Essences AR application to scan the bottles of the Beach Plastic package. After scanning, they will be able to learn and follow the journey behind the company's process of bottles made out of plastics that were found and picked up from some of the dirtiest beaches. It also shows a video to the customers where the Herbal Essence bottle is surrounded by beach scenery while at the same time displaying an informative video about the product and the plastic waste issue. At the end of the AR experience, users are shown plastic waste washing up on the beach shore and can easily swipe the screen to help clean up the beach (Wikitude)	
Stage	In market	
Beneficiaries	Stakeholders	• Product packaging plays an essential role in the consumer decision and is the ideal opportunity to share not only information about the product itself but also company values • Raising awareness about recycling and its real importance in eliminating plastic waste
	Environment	• Nearly three tons of waste from beaches(with TerraCycle) • Conservation of natural resources (crude oil) • Reduced environmental impact on marine life • Better energy saved • Fewer emissions generated

Case Study 10

Precision Maintenance

- Countries of implementation: Global
- Company: Regal Beloit Corporation

Regal Beloit Corporation was founded in 1955; it is the leading producer of electrical motion controls, electric motors, power transmission products, and power generation that provide services for markets worldwide. This company works in three different sectors: commercial and industrial systems, power transmission solutions, and finally climate solutions (Regal Beloit).

Technology	Internet of things (IoT)/augmented reality (AR)
Description	This technology can be used on tablets, mobiles, or wearable glasses; it is used to be able to monitor machines in industries that are used for packaging and to be able to get real-time data of the machine's information. The user starts by scanning a physical mark on the machine, and then by using this technology, users will be able to know the condition of the machine by reading temperature, vibrations, and oil condition, speed, and dashboards showing the overall health of machine performance. Workers can walk around the machine and collect real-time data values tied to the physical component and time series charts for each data value with inspection instructions. This technology allows workers to immediately recognize any machine faults or degradation which will greatly prevent any future machinery failure (Regal Beloit)
Stage	In market

Beneficiaries	Employees	• Enhanced data visualization • Improved employee training • Enabled precision maintenance activities • Safer work environment • Decreased operational interruptions • Reduced time to diagnose issues, communicate, and schedule corrective actions
	Companies	• Increased proactivity to asset manage • Reduced unnecessary expenses associated with equipment degradation, unplanned failures, and ineffective maintenance strategies • Reduced unplanned downtime • Improved spare parts managements

Case Study 11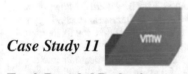

Track Recycled Packaging

- Countries of implementation: Global
- Company: VMware

VMware is an innovation company founded by Diane Greene, Mendel Rosenblum, Scott Devine, Ellen Wang, and Edouard Bugnionin in 1998. The company aims to provide mobility, security, and networking offerings to form a digital foundation for applications and services; it uses technologies, such as artificial intelligence, block-chain, machine learning, and others to solve customer's challenges. This company is transforming entire industries using its technologies, including health care, banking, manufacturing, transportation, and others.

Technology	Bloc/digital technology
Description	VMware uses blockchain technology to allow users to trace and track recycled plastics. Dell uses VMware's technology and in 2014 they started using plastics recovered from e-waste and recycled them. Hence, Dell uses this technology so that customers can see the origin of the recycled material and track its journey. Dell has a supply chain that consists of companies that collect plastic, recyclers who process waste, and manufacturers who use the recycled materials into making or producing packaging for Dell laptops. This blockchain technology supports all of these transactions that occur between companies, recyclers, and manufacturers. Therefore, Dell uses this technology to prove the reliability of these recycled materials. Moreover, this technology also provides Dell with verification of whether individuals who collect plastics are getting paid or are exposed to forced labor by working for someone for free (Otieno 2019)
Stage	In market
Beneficiaries	Stakeholders	• Track and trace recycled plastic • Improved communication and real-time monitoring of the waste materials • Improved image of Dell • Environmentally friendly • Demonstrated authenticity of reused material

Case Study 12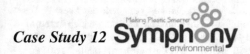

SYM Fresh Packaging

- Country of implementation: UK
- Company: Symphony Environmental Plc

Symphony Environmental Plc is a global company founded in the year 1995 with over 70 distributors worldwide where it aims to develop plastic biodegradable solutions and design technologies to change and improve normal plastic products into biodegradable material. Products are now being designed with Symphony Environmental's plastic in nearly 100 countries (Symphony Environmental 2020).

Technology	Green technologies/biodegradable or protective technology

Description	SYM Fresh bags are designed and produced with d2p ethylene absorber technology which absorbs and prevents ethylene from escaping which in return manages the moisture of the bag and hence controls the drying out process. As a result, the product's shelf life will increase, and the nutritional qualities of the fruits and vegetables will be protected and preserved; therefore, the flavor will be protected as well (Symphony Environmental 2020)	
Stage	In market	
Beneficiaries	Manufacturers	• Improved protection against bacteria, fungi, molds, mildew, and algae for food products • Increased shelf life of fruits and vegetables • Recyclable packaging • Management of waste
	Environment	• Decreased pollution • Reduced carbon emission

Case Study 13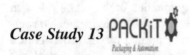

Automatic Bagging Systems

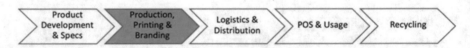

- Country of implementation: USA
- Company: PACKiT Systems

PACKiT is an Indian company that tries to find innovative solutions to make use of technologies, such as artificial intelligence and the internet of things to reduce labor and human error. The company aims to increase efficiency and productivity (PACKiT Systems).

Technology	Artificial intelligence (AI)/robotics technology
Description	PACKiT developed a machine to provide robotic bagging systems where it will automatically bag the products and it includes packages in different sizes, and depending on the size of the product, it automatically uses a certain bag size. The bag's sizes vary from 200 mm to 500 mm in width and from 350 mm to 700 mm in length. It improves productivity greatly compared to humans as it has a speed of approximately 5 seconds per packaging order; therefore, it can replace the work of eight workers. Moreover, this automated machine also prints out high-resolution texts, graphics, and barcodes which are directly pasted on the outside of the bag. This machine is offered to many industries to reduce labor and human error (Packit Systems)
Stage	In market
Beneficiaries	Stakeholders

		• Decreased human error • Lowered labor costs • Improved safety • Improved productivity • Increased speed • Printable texts and barcodes on the outside of the bag eliminating the need for a different machine

Case Study 14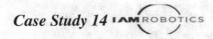

SWIFT™

- Country of implementation: USA
- Company: IAM Robotics

IAM Robotics is an autonomous robotics company founded by Tom Galluzzo in 2012. It aims to develop technologies and find innovative solutions to transform the material handling industry (IAM Robotics).

Technology	Artificial intelligence (AI)/robotics technology and sensors
Description	The swift robot is an autonomous robot that is designed for material handling. It is designed using sensors and artificial intelligence to be able to move around freely while detecting obstacles in their way and to see and locate objects in 3D and real time. It can navigate its way through different aisles using obstacle detection technology and sensors. This robot machine can pick up the package from the shelf and transport it to the desired destination at human speeds. Swift has an articulated arm which is one of the fastest industrial arms designed in the market; therefore, it can pick packages very quickly and accurately from both sides of the aisle. Moreover, it can pick up packages that are located 7 feet high on shelves and weigh up to 15 pounds. By the end of 2018, IAM Robotics was able to raise more than $20 million in funding (IAM Robotics)
Stage	In market

Beneficiaries	Packaging industries	• Packages transported quickly and with increased accuracy • Decreased possibility of human error • Decreased labor costs • Ability to heavy lift • Works independently or collaboratively • Improved speed of operation

Case Study 15 ✿PERUZA

Deposit-Refund System

Product Development & Specs > Production, Printing & Branding > Logistics & Distribution > POS & Usage > Recycling

- Country of implementation: Latvia
- Company: Peruza

Peruza is an engineering and manufacturing company; it increases the quality of the product at less cost. It aims to use artificial intelligence to find innovative solutions for other companies and to also increase efficiency (Peruza).

Technology	Artificial intelligence (AI)/sensors	
Description	There are two main problems regarding plastic packaging in the world; one is a reduction of the volume of production of plastic and two is collecting plastic effectively. Therefore, this company has created a technology using artificial intelligence with sensors embedded in it to collect plastic packaging. It can recognize and collect different types of packaging. It can accept new types of packaging (not just a standard type) as soon as the artificial intelligence can recognize it using a certain barcode, recognition of the image on the package, the shape of the package, the weight, and the spectroscopy of the material. This machine can recognize and collect plastic bottles, aluminum cans, glass bottles, jars, PET bottles, and cardboard (Peruza)	
Stage	In market	
Beneficiaries	Environment	• Variety of containers accepted • Reduced waste in ocean and streets
	Staff	• Effective time management • Easier for staff to use and clean
	Users	• This technology can be available at mass attraction events including festivals and parades • Users will enjoy themselves and will also feel like they contributed to the environment

Case Study 16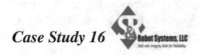

Palletizer and De-Palletizer

- Country of implementation: USA
- Company: S&R Robot Systems LLC

S&R Robot Systems LLC is a robotic automation company that provides and designs autonomous solutions for other businesses and companies to increase efficiency, enhance safety, enhance productivity and quality, and reduce expenses. Some of the services that S&R Robot Systems LLC provides are consultation, robot systems design, maintenance, repair, and industrial automation (S&R Robot Systems LLC).

Technology	Artificial intelligence (AI)/robotic technology
Description	S&R Robot Systems LLC has designed a robotic technology that uses its arm for palletizing and de-palletizing boxes and packages. This technology can check weights of packages, shrink wrapping, read barcodes, and act as a pallet and partition dispenser. Using this technology there will be more flexibility in packaging and quicker product development. Moreover, this technology can pack using cartons, cases, bags, kegs, barrels, and boxes. Depending on the tasks needed, the robot can perform other tasks, such as pallet placement, labeling of containers, and sheet placement (S&R Robot Systems LLC)
Stage	In market
Beneficiaries	Workers

Beneficiaries	Workers	• Improved time management • Optimum efficiency • Decreased maintenance needed

Case Study 17 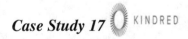 KINDRED

Piece Picking Robot

- Country of implementation: USA
- Company: Kindred AI

Kindred AI is an autonomous robotics company founded by George Babu, James Bergstra, and Suzanne Gildert in 2014. It aims to design machines using artificial intelligence to interact in the physical world and solve real-life problems (Crunch base).

Technology	Artificial intelligence (AI)/robotics technology
Description	The Kindred Sort is a robotic technology designed by Kindred AI using artificial intelligence. So far, this robot is being controlled by six pilots using Kindred's software alongside a keyboard and a 3D mouse. Using the 3D mouse, they can control and operate the robot's arm either in the vertical direction or the horizontal direction. The company's robotic arm up until now is only designed to do specific tasks, such as sorting products, grasping one item, and scanning its barcode then sending it to the final destination. It can also identify different items, differentiate between them, and send them to the correct end pile (Statt 2017)
Stage	In market

Beneficiaries	Workers	• Adaptable to changing environments • Increased product capacity • Reduced maintenance costs • High quality of work • Minimal training required

Case Study 18 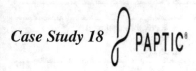 PAPTIC®

Tringa

- Countries of implementation: Global
- Company: Paptic

Paptic is a manufacturing company founded by Esa Torniainen and Tuomas Mustonen in 2015. It aims to produce high-quality recyclable and renewable material that can be used instead of plastic to enhance the environment.

Technology	Green technologies/wood fiber
Description	Paptic has created Tringa which is a wood fiber material and is completely free from paper, nonwoven, and plastic materials. It can be recycled and reused multiple times; therefore, it is a very sustainable packaging solution and eliminates single-use plastic packaging. It is also very flexible (can be fitted into a purse or pocket) and strong, hence can carry heavy products inside it; it can be ordered in any size ranging from 45 gsm (grams per square meter) up to 140 gsm. This material is already in use in different applications, such as grocery bags, boutique bags, food service bags, shoe bags, silk paper, gift wrap, garment bags, and many more (Paptic)

Stage	In market	
Beneficiaries	Customers	• Reusable material • Flexible • Strong
	Environment	• Recyclable • Renewable

Case Study 19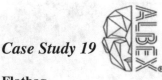

Flatbag

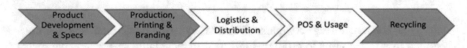

- Countries of implementation: Global
- Company: Albex DXB

Albex DXB is a company that specializes in fashion, retail, and packaging. They try to find innovative solutions to help the environment and add value to the brand; also, Albex provides advice and perspective to other businesses to help restore their brand name and bring their ideas to life (Albex DXB).

Technology	Green technologies/recyclable	
Description	Flatbag was designed in 2009 using a robotic machine where only 50% of glue was used as it was only applied where it was necessary to provide resistance. It was designed using a low gram paper that is completely free from plastic and thus the environment can be protected; it is biodegradable. This bag uses only 50% glue, saved 30% of trees from deforestation, saved 30% of paper from being wasted, and saved 30–40% of carbon emission. The weight of the bag is very light and is about 70 gsm (grams per square meter); therefore, as the number of bags increase, over 30% of the supply chain costs are saved by optimizing the size. Almost 330,000 flat bags can be transported in 1 truck whereas only 264,000 traditional bags can be transported. Moreover, it can carry products up to 10 kg and can be decomposed within 60 days, hence reducing the effect on the environment (Albex DXB)	
Stage	In market	
Beneficiaries	Environment	• Biodegradable • No plastic used • Saved 30–40% of carbon emission • Saved 30% of trees from deforestation • Saved 30% of paper from being wasted • Low weight bag
	Businesses	• Ability to take weight up to 10 kg • Used in many different sectors • Space-saving • Environmentally friendly

Case Study 20

Mountain Dew Kickstart VR APP

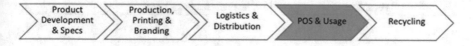

- Country of implementation: UAE
- Company: Hive Innovative Group

Hive Innovative Group is a digital advertising agency founded in 2010. Hive aims to combine all sectors under one digital communication platform, including mobile designers, innovation experts, brand strategists, and VR or AR specialists. By 2012, the Hive Innovative Group had become the first communication agency to grow into the digital market.

Technology	Artificial intelligence (AI)/virtual reality (VR)	
Description	Hive Innovative Group has come up with a VR experience called "Feel the Power" to help PepsiCo after launching its new product Mountain Dew "Kickstart." During this experience, the users wear VR glasses and feel the energy and the power of the product in a ski diving experience where they can see all the different packaging of the kickstart drink which makes them want to continue buying the product. Users can fly, watch the colorful cans that are surrounding them in the air, jump around them, and hang onto them. The main goal of this application is to make users want to buy the product after seeing all the colorful cans flying in the air (Hive Innovative Group)	
Stage	In market	
Beneficiaries	Customers	• New product awareness • Extraordinary energetic experience via VR tool
	PepsiCo	• Customer feedback data

Real Estate

Abstract The term *real estate* means real or physical property and its industry rotates around land ownership, buildings, air rights (above the land), and underground rights (below the land). This industry's business model varies from country to country, due to its direct relation with regulatory policies, laws, and culture. The duration of occupation is one of the longest business cycles due to the length of time it takes to build and then occupy properties.

Keywords Resource optimization · Fully utilized properties · Optimized buildings · Chatbots · Virtual reality touring · Immersive experience · Consumption optimization · Maintenance detection · Irrigation optimization · Social media listening · Infrastructure development · Centralized platforms · Centralized data · Customer experience · Big data analytics

The term *real estate* means real or physical property and its industry rotates around land ownership, buildings, air rights (above the land), and underground rights (below the land). This industry's business model varies from country to country, due to its direct relation with regulatory policies, laws, and culture. The duration of occupation is one of the longest business cycles due to the length of time it takes to build and then occupy properties.

One of the most prominent challenges this industry is faced with is poor infrastructure. A lot of countries do not maintain and repair their bridges, roads, and buildings. Unfortunately, any real estate property in an area with poor infrastructure is usually in low demand as, most often, people do not want to deal with deteriorated groundwork and frail environments. Technology can help this industry by allowing buyers and sellers to communicate directly together, and view locations and properties virtually,

while eliminating costly middle agents (linchpin seo 2020).

The market is not stable and noteworthy trends are taking place (Finance Monthly, 2020). In Egypt, real estate has multiplied cost-wise in the past decade, due to the government's strategic change in new city development of selling large plots to developers and allowing them to deliver the products or services directly to their end users. This has increased competition in the local market.

© Springer Nature Switzerland AG 2023
M. Anis et al., *Mapping Innovation*,
https://doi.org/10.1007/978-3-030-93627-3_11

Value Chain

The real estate value chain is made up of five phases: holding and ownership, marketing, construction, transactions, and facilities.

Holding and ownership refer to defining a construction project and its ownership, which usually involves real estate developers and investors. The real estate is then evaluated for pricing and promoted to the appropriate audience during the marketing phase. Construction then begins and includes the procurement of required materials, contract exchanges, and man aging the workforce involved.

Once the real estate is ready then brokers and agents are contracted to sell the real estate to customers during the transaction phase. Finally, facilities may be offered via external third parties to improve the value of the real estate being sold (facilities phase). These facilities may include but are not limited to surveillance, security, maintenance, and entertainment.

Case Studies

Case Study 1 PointGrab

Machine Learning in Building Automation Systems

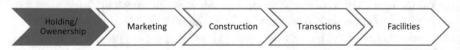

- Country of implementation: Israel
- Company: PointGrab

PointGrab is a pioneer company in the field of workspace optimization solutions that uses embedded analytics sensors to gather information about the behavior of real estate spaces and therefore utilize its resources in real time.

Technology	Artificial intelligence (AI) and internet of things (IoT)/machine learning
Description	PointGrab uses artificial intelligence (AI) in real estate to optimize the different resources inside the buildings. This involves putting a sensor in strategic points inside the building such as in the lighting systems, the heating systems, elevators, or security systems. These sensors shall gather the data that will help the owner to improve energy efficiency. Machine learning analytic algorithms are then applied to these data which has a large variance. This helps real estate owners to predict what will be going on with their facilities at a certain time under certain circumstances so that they can respond accordingly (Daniel 2019)
Stage	Recent to market
Beneficiaries	Residents

	• Fully utilized properties • Optimized building resources

Case Study 2 @ppfolio PROPERTY MANAGER

Automation in Property Management

- Country of implementation: USA
- Company: AppFolio

AppFolio is a company that is founded in 2006 and provides software services applications for property management.

Technology	Automation and digitalization	
Description	If managers were able to fully view the sales and agent's performances, they can drive prospect conversions and enhance their business outcomes. AppFolio developed real estate chatbots that interact with customers, through Amazon Echo or other smart devices, to support all related services. They also accommodate applications for potential tenants by appointing viewings, through a virtual conversation between the users and a chatbot, giving the impression of real agents (Daniel 2019)	
Stage	In market	
Beneficiaries	Buyers	• View real estate virtually • Chat with bots 24/7

Case Study 3

- Country of implementation: USA
- Company: Airbnb

Airbnb Inc is a rental online marketplace company that provides arrangements for primary homestays and tourism experiences (Matellio 2020).

Technology	Platforms	
Description	Airbnb provides a platform that uses these recommendations to match tenants with potential landlords (Daniel 2019)	
Stage	In market	
Beneficiaries	Tenants and landlords	• Matching tenants with suitable properties • Cost-effective • Satisfaction for both parties

Case Study 4 matterport

Virtual Property Showcase

- Country of implementation: Online platform
- Company: Matterport

Matterport is a 3D data platform company that provides virtual tours for real estate agents using their 3D cameras and interactive viewing platforms. They allow users to capture and upload digital versions of real-world environments and share them with others.

Technology	Virtual reality (VR)
Description	Matterport uses virtual reality (VR) technology to allow people to virtually visit places. Clients put on a virtual reality (VR) headset and then can "walk through" properties in an immersive three-dimensional experience. Potential buyers can view many properties quickly and conveniently (Gleb 2020)
Stage	Recent to market
Beneficiaries	Landlords

	Landlords	• Capture and upload properties • Reach more potential buyers • Increased sales
	Buyers and tenants	• 24/7 virtual tours • Time-effective • Cost-effective • Global access

Case Study 5 rooomy

Virtual Staging

- Countries: Netherlands and China
- Company: RoOomy

RoOomy is a leader company in 3D modeling, virtual staging, and augmented reality app development. They provide their services in real estate and home furnishing.

Technology	Virtual reality (VR)/augmented reality	
Description	RoOomy uses augmented reality to make properties look homely without real furniture. According to a staging report done by the national association of realtors, 77% of real estate agents say that staging helps clients associate a place with their future house (Gleb 2020)	
Stage	In market	
Beneficiaries	Buyers and tenants	• Bespoke property visualization
	Landlords	• Eliminated need to stage properties • Cost-effective • Time-effective

Case Study 6

Big Data in Real Estate

- Country of implementation: USA
- Company: Zillow

Zillow Group Inc. is an American real estate company that started as a media company by selling ads on its website. Afterward, Zillow began to work in real estate search engines by engaging with more than 180 US newspapers.

Technology	Big data analytics	
Description	Zillow provides a website that caused significant improvement by using big data. They collect and compile data on selling and buying trends that are related to demographic information and then analyze this to offer insights on various parameters, such as pricing and home value trends (Bernards 2020)	
Stage	In market	
Beneficiaries	Buyers and tenants	• Access to multiple properties • Price comparison
	Zillow group	• Revenue source (Ads)

Case Study 7

Internet of Things (IoT) Meters for Monitoring

Holding / Owenership → Marketing → Construction → Transctions → Facilities

- Country of implementation: Denmark
- Company: Develco Products

Develco Products is a business-to-business company that provides white label products in the field of smart homes and building management.

Technology	Artificial intelligence (AI)/smart metering	
Description	For a sustainable, future-proofed community, Develco Products use smart internet of things (IoT) meters to allow residences to monitor and calculate consumption. Using this technology benefits both the community management company as well as the residence. For community management, they can see in a much clearer way in a dashboard format the consumption levels and the distribution of the consumption and the consumption development in the community through the years. This will enhance and optimize the way developers design the infrastructure in the community. Furthermore, from the resident side, it gives a clearer idea about their consumption and provides a dashboard that can be customized per room so that they can see the consumption of each room on its own for better optimization (Develco Products 2020)	
Stage	In market	
Beneficiaries	Developers	• Live consumption dashboard • Better optimization • Better monitoring • Easier money collection
	Residents	• Improved understanding • Home dashboard • Seeing consumption per space • Warning notifications (over usage)
	Utility companies	• Easier and faster collection of money • Optimized consumption

Case Study 8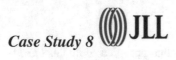

Internet of Things (IoT) in Predictive Maintenance

| Holding / Owenership | Marketing | Construction | Transctions | Facilities |

- Country of implementation: USA
- Company: JLI

JLI is a world pioneer company in real estate services as they buy, build, and invest in a variety of commercial real estate assets.

Technology	Artificial intelligence (AI)/predictive maintenance	
Description	JLI is using internet of things (IoT) sensors to detect faults before they happen. Using internet of things (IoT) sensors sends a signal with the nature of the defects and the exact location of any defects. This is extremely crucial because the maintenance team will be able to fix the issue without the need of waiting and seeing the impact of the defect (Penny 2018)	
Stage	In market	
Beneficiaries	Residents	• Hassle-free living
	Buildings	• Increased life span • Maximum efficiency

Case Study 9

Smart Irrigation

| Holding / Owenership | Marketing | Construction | Transctions | Facilities |

- Country of implementation: Belarus
- Company: Digiteum

Digiteum is a company that develops multifunctional software focusing on designing digital tools. Digiteum is working based on a result-oriented approach that ensures successful outcomes as soon as possible.

Technology	Artificial intelligence (AI)/smart irrigation
Description	One of the most water-consuming elements in the real estate industry is irrigation. Digiteum uses internet of things (IoT) sensors that detect the moisturization levels in the soil and in the roots of the plants to signal a water supply to automatically turn on and off. These sensors can be programmed with the plant's characteristics and special treatment at different times and in different conditions (Digiteum 2019)
Stage	Recent to market

Beneficiaries	Residents	• Maintained beauty of the landscape • Minimal water usage • Cost-efficient
	Developers	• Reduced maintenance fees • Pleasant landscaped environments
	Plants	• Improved plant life

Case Study 10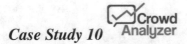

Understanding People on Social Media

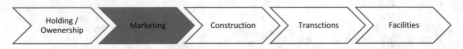

Holding / Owenership → Marketing → Construction → Transctions → Facilities

- Region: MENA Region
- Company: Crowd Analyzer

Crowd Analyzer uses technology to take the lead in developing the applications of artificial intelligence (AI) and machine learning toward natural language processing inside the MENA region.

Technology	Big data	
Description	Crowd Analyzer is specialized in Arabic-based Artificial Intelligence (AI) on social media platforms. It detects humor, sarcasm, and slang, which has a great impact on real estate where developers can better understand the needs and the feedback of the customers (Crowd Analyzer 2020)	
Stage	In market	
Beneficiaries	Developers	• Useful data • Evaluate launched projects and features • Enhanced sales performance • Better designed products
	Social media platforms	• Better monetarize data
	Clients	• Access to higher quality products • Share opinions and concerns

Case Study 11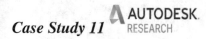

Big Data and Efficient Designs

- Country of implementation: USA
- Company: Autodesk Research

Autodesk designs and develops software for people with explorations over several key areas specifically the architecture industry and artificial intelligence (AI).

Technology	Big data/smart program allocation	
Description	Autodesk Research uses big data so that computers can learn from previously made designs to create an endless number of iterations to achieve the perfect design with minimum waste potential (Mehdi 2017)	
Stage	In market	
Beneficiaries	Developers	• Bespoke designs • Maximized revenue • Faster sales (quality design) • Faster design process (faster cash flow)
	Designers	• Endless design options • Optimized designs • Faster design processes
	Tenants	• Cost-effective • Improved sustainable development (higher utilization percentage)

Case Study 12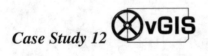

Augmented Reality in Real Estate

Holding / Owenership — Marketing — Construction — Transctions — Facilities

- Country of implementation: Canada
- Company: vGIS

vGIS is a company that develops visualization platforms to transforms traditional BIM, CAD, and other types of data into augmented reality visuals.

Technology	Augmented reality (AR)	
Description	vGIS uses augmented reality (AR) in an extremely useful way to project the infrastructure design over the physical space to pinpoint faults thus saving time and money (XYIIT 2020)	
Stage	In market	
Beneficiaries	Developers	• Cost-effective • Time-effective • Promoting high-tech orientation • Improved reputation • Premium pricing
	Contractors	• Optimized investigation • Improved communication
	Maintenance team	• Improved reporting and fixing of defects

Case Study 13   **LUNAS˙**

Virtual Reality (VR) Touring in Real Estate

| Holding / Owenership | Marketing | Construction | Transctions | Facilities |

- Countries of implementation: Canada and Belarus
- Company: Lunas 3D

Lunas 3D is a visualization company that consists of hand-picked computer graphics artists with solid professional experience. They blend architectural designs with an artistic taste to render 3D VR visualization of real estate and construction assets.

Technology	Virtual reality (VR)	
Description	Lunas 3D uses virtual reality (VR) technology to allow developers to create walkthroughs using the mobile phone app and virtual reality (VR) glasses (Lunas 3D 2020)	
Stage	In market	
Beneficiaries	Developers	• More informative tools • Impressive presentations • Positive word of mouth • Potentially increased sales • Increased revenue
	Clients	• More immersive experience • Ability to see the minor details
	Salespersons	• Higher closing ratio • Powerful sales aids

Case Study 14

AUTODESK.
RESEARCH

Virtual Reality (VR) and Architectural Designs

| Holding / Owenership | Marketing | **Construction** | Transctions | Facilities |

- Country of implementation: USA
- Company: Autodesk

Autodesk designs and develops software for people with explorations over several key areas.

Technology	Virtual reality (VR)	
Description	Autodesk uses a virtual reality (VR) application so that clients can visualize and choose the design options they prefer, in detail, which match an actual finish quality (Redshift Autodesk 2020)	
Stage	In market	
Beneficiaries	Designers	• Express designs more interactively • Limitless design options • Minimized rework and resubmission
	Developers	• Better communication • Positive word of mouth
	Clients	• Accurate presentation • Ability to see the design before execution

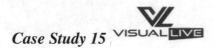

Case Study 15

Augmented Reality (AR) in Construction Sites

| Holding / Owenership | Marketing | Construction | Transctions | Facilities |

- Country of implementation: USA
- Company: VisualLive

VisualLive is an augmented reality (AR) software-as-a-service company that works on saving time by connecting the field and office.

Technology	Augmented reality (AR)	
Description	By utilizing the latest technology in augmented reality (AR), VisualLive allows them to see the site progress in a more immersive way. This enhances the buying experience and builds a healthy relationship between developers and property buyers (Peter 2019)	
Stage	In market	
Beneficiaries	Developers	• Building credibility • Improved communication • Positive word of mouth
	Buyers/clients	• Immersive experience • Access to sits updates 24/7

Case Study 16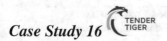

Tender Platforms for Contractors

- Country of implementation: Online Platform
- Company: Tender Tiger

Tender Tiger is an online platform that tracks more than two million tenders annually whereas subscribers have access to all relevant notices in the categories they are interested in.

Technology	Platforms
Description	Tender Tiger has created a platform that organizes and regulates the relationship between the developer and the different contractors which also allows developers to subcontract, in a way that would not have been possible previously. This ensures that the developers deliver the project at the lowest cost (Tender Tiger 2020)
Stage	In market

Beneficiaries	Contractors	• More transparency • Better collaboration
	Developers	• Cost-effective • Time-effective • Organized process
	Project managers	• More control

Case Study 17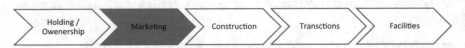

Real Estate E-commerce

Holding / Owenership	Marketing	Construction	Transctions	Facilities

- Country of implementation: Egypt
- Company: SODIC

SODIC has two decades of successful real estate operations in Egypt. Based in Cairo, SODIC is listed on the Egyptian stock exchange as it brings to the market award-winning developments that meet Egypt's ever-growing need for high-quality housing.

Technology	Platforms

Description	With the expansion of the real estate market, the number of new properties for sale is on the rise. For property buyers, the choices are limitless and with that emerges the need for a centralized platform that acts as a marketplace, allowing shoppers to look for their dream property. SODIC developed a marketplace that facilitates products to a wider customer base who are actively looking for properties. This allows developers to create digital launches, as this marketplace has the potential of becoming a digital broker to all property buyers (SODIC 2020)	
Stage	In market	
Beneficiaries	Buyers	• Wider variety of offering • Ability to leave reviews and feedback • More transparency • Ability to compare and evaluate offers
	Developers	• Wider customer base • Presentation channels to showcase projects • Automatic analytics tool
	Brokers	• Centralized data and information center • Business account (consulting services)

Retail

Abstract The retail industry is the process of which goods or commodities are purchased from manufacturers, then providing those goods for sale to consumers at a marked-up price and offering services in support of those products. The industry's main activities include managing how merchandise is ordered, purchased, allocated, received, stored, and distributed to stores or directly to consumers in a timely and accurate manner. Additionally, the industry focuses on processes associated with keeping stores operational and profitable, serving customers, and selling products or services over the internet. Communicating with customers to persuade them to buy products or services and sustaining the relationship between the retail enterprise and customers to drive customer loyalty are essential aspects of the retail business. This requires identifying desired services and products and communicating those via customers' preferred channels (An Overview of Retailing).

Keywords AR visualization · Virtual placement · Radio frequency identification · In-store experience · AI-based robots · Digital authentication · Quality optimization · AR tracking · Chatbot · Demand prediction · Virtual assistant · AI-based drone · Customized shopping experience · Automated storage · Automated delivery · Safe financial transactions · Self-checkout

The retail industry is the process of which goods or commodities are purchased from manufacturers, then providing those goods for sale to consumers at a marked-up price and offering services in support of those products. The industry's main activities include managing how merchandise is ordered, purchased, allocated, received, stored, and distributed to stores or directly to consumers in a timely and accurate manner. Additionally, the industry focuses on processes associated with keeping stores operational and profitable, serving customers, and selling products or services over the internet. Communicating with customers to persuade them to buy products or services and sustaining the relationship between the retail enterprise and customers to drive customer loyalty are essential aspects of the retail business. This requires identifying desired services and products and communicating those via customers' preferred channels (An Overview of Retailing).

The retail industry faces a lot of challenges and aims to find solutions to those adversities. First of many is the demographical shift. Millennial consumers who have grown up using the internet now hold buying power. It is evident that online shopping has become more popular and is in demand. Retailers must learn how to engage this demographic through products and shopping experiences. At the same time, they still need to retain the older generations who still hold buying power but may have different retail tastes and preferences.

Additionally, there has been an increase in dependence on devices. The physical and digital worlds have converged and will continue to emerge moving forward. The idea of on-ground retailing is coming to an end. Another challenge is the uncertainty about the future. The world's economy is fluctuating in nature, and it became vital for retail businesses to predict the market and consumer trends to have a competitive edge.

Even though the retail sector is faced with challenges, it is also manifested with opportunities. With the rise in technology, businesses can improve customer experience by providing more personalized products or services and marketing strategies. Furthermore, technological advancements in big data analytics have helped retail companies manage large amounts of data, which may help predict trends and increase efficiencies.

There has been a widespread shutdown of stores in apparel, specialty, and department store segments as a shift to online and digital purchasing are increasing and may influence the growth of many businesses as creating online platforms is easy and is a lot more convenient for customers. Customers can purchase any commodities at any time from anywhere, which may help increase business revenue.

Value Chain

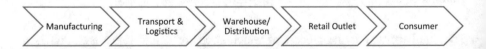

The value chain of the retail industry is characterized by five main activities: manufacturing, transport and logistics, warehousing/distribution, retail outlet, and consumer.

In the first phase of the chain, different industries such as food and beverages, clothing, pharmaceuticals, and many others process inputs and package them in ways that are useful and attractive to consumers. During the next phase, transportation and logistics, the merchandise is transported to warehouses for distribution (warehousing/distribution phase).

From the warehouses, the merchandise is sent to retail outlets (retail outlet phase) to be finally purchased by the consumers (consumer phase).

Case Studies

Case Study 1

Augmented Reality (AR) in Retail

- Countries of implementation: Global
- Company: IKEA

IKEA is a multinational furniture retail company specializing in designing and selling furniture, home appliances, accessories, and useful home commodities. The company's primary goal is to create a better home through high-quality and affordable home goods (IKEA | Wikipedia).

Technology	Augmented reality (AR)
Description	IKEA has launched an application called IKEA Place that uses augmented reality technology to allow consumers to experiment and visualize how a particular commodity looks in their homes before buying it. The app scales the product based on the room's proportions with high accuracy. It allows users to visualize and try different products with as many styles, shapes, and colors through their smartphones. The application also enables users to share an image of the augmented setup with friends and family (IKEA Launches IKEA Place, a New App That Allows People to Virtually Place Furniture in Their Homes | IKEA 2017)

Stage	In market	
Beneficiaries	Consumers	• Easier and increased confidence in decision-making • Prepurchasing product visualization • Ability to share visualization with family and friends

Case Study 2

Internet of Things (IoT) Appearance in Retail

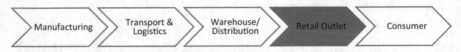

Manufacturing → Transport & Logistics → Warehouse/Distribution → Retail Outlet → Consumer

- Country of implementation: USA
- Company: Levi Strauss & Co.

Levi Strauss & Co. is an American-based company that mainly does cloth manufacturing. It is internationally known for its denim jeans Levi's. The company now manufactures all kinds of jeans and pants as well as different clothing items such as eyewear, hats, sweaters, shirts, and more (Levi Strauss & Co. | Britannica).

Technology	Internet of things (IoT)/radio frequency and cloud technology
Description	Levi is planning to use radio frequency identification (RFID) tags to overcome the challenges retail stores face with misplaced items. As the company tries to enhance consumers' in-store experiences, it decided to use intel's powered sensors and cloud-based analytics to enable customers to find the clothing they desire and help minimize the chances of missing good deals due to missing items. Finding missing items in retail clothing stores are essential, especially if it was a unique good. Thus, RFID tags are placed on all items in the store, and the RFID antennas are always on. The intel system can scan every product and locate it. Additionally, the system is embedded with an alarm to alert employees when the stock of any item is low (Levi's Real-Time Tracking of Jeans: RFID in Retail I RTInsights 2016)
Stage	Testing phase

Beneficiaries	Consumers	• Alerts • Ability to easily find product • Consistent product inventory

Case Study 3 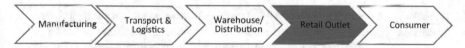 CaliBurger

Transforming Retail Pain into Smart Gains

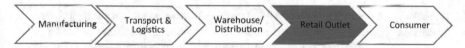

Manufacturing › Transport & Logistics › Warehouse/ Distribution › Retail Outlet › Consumer

- Country of implementation: USA
- Company: CaliBurger

- CaliBurger is a restaurant in Los Angeles that produces and sells fresh burgers in California style (About Us I CaliBurger).

Technology	Artificial intelligence (AI) and machine learning/automation and digitalization	
Description	Flipping burgers in the harsh environment of the kitchen is associated with common risks. Not everyone can cope with hot oil, burns, and cuts in this work field. The fast-food businesses, especially the burger chain, are high in demand, and the industry is trying to find solutions to fulfill needs while minimizing hazards in the kitchen. Miso Robotics has created an AI-based automated robot that flips burgers called Flippy. The robot can complete the tedious, harmful, and dull work to cook burgers. CaliBurger is the first burger chain restaurant to use Flippy, a robot equipped with safety sensors, 3D and thermal sensors, and cameras. Flippy can take a burger patty, unwrap it, put it on the grill, and cooks it. It records the burgers' cook time and temperature and then alerts employees when the job is done (Meet Flippy, A Burger- Grilling Robot from Miso Robotics and CaliBurger I TechCrunch 2017)	
Stage	In market	
Beneficiaries	Consumers	• Improved food quality • Improved meal quality • Sterilized food preparation environment
	Burger-chain restaurants	• Improved kitchen safety • Ability to meet high demands • Increased revenue

Case Study 4

Future of Shopping

- Countries of implementation: Global
- Company: Zara

Zara is one of the world's most significant apparel retailers. The company is based in Spain and was founded in 1975. It now has over 2000 stores worldwide. The company specializes in producing trendy clothing and products, including accessories, beauty, swimwear, perfumes, and more (Zara: Company Snapshot | Business of Fashion).

Technology	Artificial intelligence (AI)
Description	Zara launched an AI-based mobile application called Zara AR that allows users to capture, using their smartphone cameras, an item in the store, and see a model wearing the selected product. They can, through the app, better understand the item in question and buy it online. This encourages people to "try on" clothing items virtually, decreasing contamination, especially during the COVID-19 pandemic (Augmented Reality + Retail: A Glimpse in the Future of Shipping | Bazaar Voice 2018)
Stage	In market

Beneficiaries	Consumers	• Reduced human contact and contamination • Quick and efficient online orders available in stores, it could easily and quickly order the item online • Enhanced shopping experience and engagement

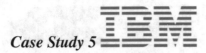

Case Study 5

The Trust Chain Initiative

- Countries of implementation: Global
- Company: IBM

IBM is an American international company that is based in New York and specializes in technology. It develops and manufactures computer hardware, middleware, and software devices and also offers technology consulting services (IBM | Wikipedia).

Technology	Blockchain
Description	IBM collaborated with the largest diamond and jewelry companies to create a blockchain platform called Trust Chain. The platform's main objective is to track and verify luxury metal and jewelry at any stage in the value chain. This IBM technological solution helps enhance the worlds' authentication of diamonds and manage its transportation safely. This blockchain technology will help provide transparency among clients and consumers securely (New Consortium Collaborates to Put Jewelry on Blockchain | Trust Chain Jewelry)
Stage	Development phase

Beneficiaries	Consumers	• Authenticity of purchases

Case Study 6 Walmart >¦<

Bringing Transparency to the Food Supply Chain

- Country of implementation: USA
- Company: Walmart

Walmart is an American retail company that functions in the chain of hypermarkets, discount stores, and grocery stores (Walmart | Wikipedia).

Technology	Blockchain
Description	Walmart uses blockchain technology to help track food in the supply chain. The company uses Hyperledger Fabric, a blockchain framework that helps create solutions through blockchain technology, to develop a system that traces its products to help quickly and more effectively eliminate any infected produce. The company tested the system on mangos in their US stores. This allows the company to trace its point of origin and its journey from start to finish (Case Study: How Walmart Brought Unprecedented Transparency to the Food Supply Chain with Hyperledger Fabric I Hyperledger)
Stage	Testing phase
Beneficiaries	Consumers

Consumers	• Improved food safety
Employees	• Quicker trace and track to identify foodborne challenges

Case Study 7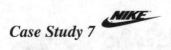

Nike Automates Process for Customer-Designed Sneakers

- Countries of implementation: Global
- Company: Nike

Nike is a multinational company specializing in designing manufacturing, creating, advertising, and selling shoes, equipment, and accessories. The company's primary goal is to develop innovative sports solutions and products to help enhance community potential (About Nike I Nike).

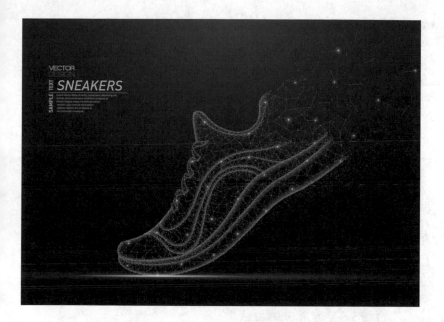

Technology	Artificial intelligence (AI)
Description	Nike has developed an AI platform to create personalized sneakers for consumers. Customers can now go to stores, fashion their shoes, and leave with them instantly. The AI platform is called the Nike Maker Experience. Customers put on Nike Presto X footwear that does not have any designs, and with the help of voice recognition software, they can select different designs and colors for their shoes. The system is composed of AR and object tracking algorithms to help demonstrate the chosen design to the client. Once the client selects the design, the system can finalize the product in just a few hours; usually, customized jobs can take weeks (The Coming AI Revolution in Retail and Consumer Products | IBM)
Stage	Testing phase
Beneficiaries	Consumers

Beneficiaries	Consumers	• Enhanced the shopping experience • Customizable shoes • Prepurchase product visualization • Fast product delivery

Case Study 8 *H&M*

Intelligent Chatbot for Boosting Retail Sales

- Country of implementation: Canada and USA
- Company: H&M

H&M is a clothing company that sells clothes compiling with the world's most recent trends. It sells clothes, accessories, shoes, bags, and more for women, men, and children. The company has over 500 stores in over 70 different countries and now is focusing on enhancing its online presence (H&M | Wikipedia).

Technology	Artificial intelligence (AI) and machine learning	
Description	H&M developed a chatbot on the messaging app Kik in 2016. This AI technology platform allows consumers to see, buy, and share products from its catalog. It mimics the job of a personal stylist as it asks questions to understand the shopper's preferences and style and uses photos to show the product. Once the app recognizes the client's profile, the customer can develop their outfits and look through outfits created by other clients or the shop that corresponds to the consumer's style (Chatbots in Retail: Nine Companies Using AI to Improve Customer Experience | Real Insight Network 2018)	
Stage	In market	
Beneficiaries	Consumers	• Enhanced customer engagement • Improved shopping experience • Personal stylist to support the shopping experience

Case Study 9

Predicting Fashion Trends

| Manufacturing | Transport & Logistics | Warehouse/ Distribution | Retail Outlet | Consumer |

- Countries of implementation: Global
- Company: Heuritech

Heuritech is a forward-looking revolutionary technology company that services companies in the fashion industry and provides them the actionable insights that are competitive and then correspond to the future dynamics of fashion and trends (Heuritech | LinkedIn).

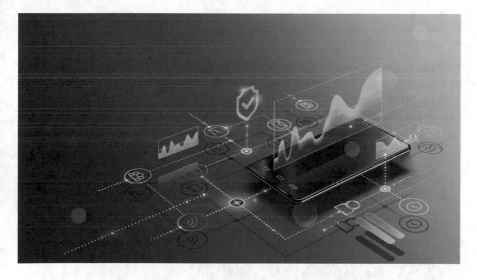

Technology	Artificial intelligence (AI)
Description	The Heuritech AI platform helps to predict trends to help in their new collections and maximize revenues. Companies tend to lose billions of dollars in unsold merchandise because they did not comply with fashion due to false predictions. With this technology, companies can predict demands early to avoid huge losses. Heuristics can analyze massive data from social media, forecast trends, and turn them into actionable insights (Trend Forecasting | Heuritech)
Stage	In market

Beneficiaries	Luxury retailers	• Accurate prediction on trends to help plan merchandise • Avoidance of loss of money due to unsold products

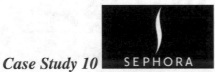

Case Study 10

AR in Makeup Application

- Countries of implementation: Global
- Company: Sephora

Sephora is a multinational company that specializes in beauty and skincare. It has many beauty stores that feature thousands of brands along its line. It has products ranging from cosmetics, skincare, perfume, nail care, hair product, and beauty equipment (Sephora | Wikipedia).

Technology	Artificial intelligence (AI)
Description	Sephora is one of the few beauty stores that offer magic mirrors and applications that help clients visualize how different cosmetics, colors, and shapes look on them using AR technology. Testing new makeup could be unclean, unsanitary, and inconvenient. Customers need to avoid sharing certain products to prevent transferring infections and diseases. Sephora has created a virtual artist who uses AI and AR technologies to enable customers to apply makeup, learn new techniques, and see how the makeup looks on them (I Learned How to Apply Makeup Using a Futuristic New Feature on Sephora's App – Here's What Happened I Business Insider 2017)
Stage	In market
Beneficiaries	Consumers

	• Prepurchase product visualization
	• Learning of new makeup techniques
	• Avoidance the use of shared products or testers

Case Study 11

Augmented Reality (AR) Transforming the Retail Industry

- Country of implementation: UK
- Company: ASOS

ASOS is an online British retail store that specializes in fashion and cosmetics. The company was founded in 200, and its target audience is mainly people in their teens and twenties (Wikipedia I ASOS).

Technology	Augmented reality (AR)	
Description	ASOS uses AR technologies in clothing to help customers see how the clothing item looks like on different body types. The app shows clients the same clothing items but on different body sizes to help adhere to the audience. This is to help accomplish the company's primary goal in spreading body positivity and fashion that includes everyone. It allows the customer to see how the product might look on a body shape similar to theirs before buying an outfit (The Four Definitive Use Cases For AR and VR in Retail | Forbes 2019)	
Stage	In market	
Beneficiaries	Consumers	• Raised awareness to create inclusive fashion • Enhanced shopping experience due to raising positive body image • Visualization of all different body types

Case Study 12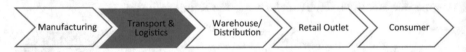

Delivery Drones

Manufacturing ⟩ Transport & Logistics ⟩ Warehouse/ Distribution ⟩ Retail Outlet ⟩ Consumer

- Country of implementation: UK
- Company: Amazon

Amazon is an American-based company specializing in technology, including AI, cloud computing, e-commerce, and more. It was first known for selling books on its website, but now Amazon has built its online presence to include products from clothes to home appliances (Amazon | Wikipedia).

Technology	Artificial intelligence (AI)	
Description	Amazon has created Prime Air, an AI-based drone used to deliver small packages in the UK. The device can make several trips in a few minutes, and the company aims to normalize this futuristic and safe delivery system. The drones are powered by AI technology and sensors to help the device fly without hitting any obstacles and eliminating threats drones might create to aircraft (Amazon Poised to Test Drone Deliveries Powered by Artificial Intelligence I TT Newsmaker 2019)	
Stage	In market	
Beneficiaries	Consumers	• Better, safer, and faster delivery

Case Study 13

Robots in Retail

- Country of implementation: Japan
- Company: SoftBank

SoftBank is a telecommunication company in Japan that was founded in 1981. The company aims to enhance its services by investing in many technological advancements in the internet of things (IoT), artificial intelligence (AI), and robotics (What Is SoftBank | Investopedia 2019).

Technology	Artificial intelligence (AI)	
Description	SoftBank, a telecommunication company in Tokyo, has been using a human-like robot to communicate with clients. The AI technology embedded in the robots helps the machine understand and analyze the customer's emotions and feelings. The company stated that this AI-platform helped enhance customer service and significantly increased sales. The robot's functionality is to connect and assist customers in the company's retail mobile stores and share any valuable knowledge with the product of interest (For Better Business Just Add Pepper| SoftBank Robots)	
Stage	In market	
Beneficiaries	Consumers	• The unique and enhanced shopping experience • Receive customized recommendations

Case Study 14

Brilliant Manufacturing

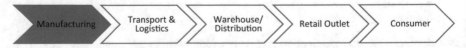

- Countries of implementation: Global
- Company: General Electric Digital (GE Digital)

General Electric Digital is an American multinational subdivision of General Electric and one of the early endorsers of big data and data analytics. The company's primary goal is to provide innovative solutions to challenge worldwide industrial companies face with software and the internet of things (IoT). It explores how assets in this industry are run, constructed, maintained, and used by machine learning, the internet of things (IoT), and big data to convert innovative insights into profitable and productive company outcomes (GE Digital | Wikipedia).

Technology	Artificial intelligence (AI) and machine learning	
Description	GE Digital has created an AI and machine learning-based platform called Proficy Manufacturing Execution System (Proficy MES). The systems provide a method to increase productivity and reduce costs for businesses during the manufacturing phase. The system is embedded with several advanced technologies that can transform analysis from data during the manufacturing process to actionable insights. Toray Plastic is one of the many companies that use GE's software to detect and eliminate faulty products (Artificial Intelligence in Retail -10 Present and Future Use Case | Emerj 2020)	
Stage	In market	
Beneficiaries	Consumers	• Have higher quality products • Reduce probability to receive defective items

Case Study 15

Automated Storage

- Country of implementation: China
- Company: Alibaba

Alibaba is a technology-based company based in China. The company specializes in the integration of technological advances and e-commerce to retail (Alibaba | Wikipedia).

Technology	Automation and digitalization/robotics
Description	Alibaba uses automated storage and retrieval system for its warehouse in China. The system is used to keep track of goods in and out of the warehouse. Skypod is one of the robots used in the retail warehouse to help retrieve the goods customer orders. This helped reduce the labor needed in the storage unit drastically (Warehouse Robotics: Everything You Need to Know in 2019 | Logiwa 2020).
Stage	In market
Beneficiaries	Employees

	• Time-effective
	• Improved labor efficiency

Case Study 16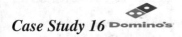

Robotic Delivery

| Manufacturing | Transport & Logistics | Warehouse/ Distribution | Retail Outlet | Consumer |

- Countries of implementation: Australia, New Zealand, Belgium, France, the Netherlands, Japan, and Germany
- Company: Domino's Pizza

Domino's is an American pizza chain restaurant that was founded in 1960. It is known as one of the leading pizza delivery companies. Its goal is to make and deliver the best pizza worldwide (What We're About: About Domino's | Domino's).

Technology	Artificial intelligence (AI)/robotics
Description	Dominos is now using a robot to do its delivery called the Domino's Robotic Unit (DRU). The DRU could deliver the food and drinks to consumers while keeping them at an optimum temperature. The machine uses technology that was used in military training as it is equipped with sensors to help it find and travel on the fastest and best path to reach the customer (Artificial Intelligence (AI) in Retail -10 Present and Future Use Case	Emerj 2020)
Stage	In market

Beneficiaries	Consumers	• Faster delivery • Appropriate food temperature and storage • Protection of products during inclement weather conditions

Case Study 17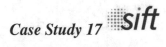

AI-Based Fraud Detection

- Country of implementation: Global
- Company: Sift Science

Sift Science is one of the leading companies in the security and digital safety. Sift's main goal is to help companies and enterprises avoid any fraud. Leading platforms like Twitter and Airbnb use Sift to help stay secure and competitive (Sift | LinkedIn).

Technology	Artificial intelligent (AI) and machine learning	
Description	One of the companies' most significant challenges is dealing with payment fraud. Clients tend to be skeptical of online payment services as they are not sure how secure the platform is. Sift Science is a company that uses artificial intelligence (AI) and machine learning-based technologies to help companies identify payment fraud. This is a significant application in the retail industry, especially considering the current growth and popularity of e-commerce (Artificial Intelligence in Retail -10 Present and Future Use Case | Emerj 2020)	
Stage	In market	
Beneficiaries	Consumers	• Safer financial transactions

Case Study 18

Checkoutless Stores

- Country of implementation: USA
- Company: Sam's Club

Sam's Club is an American- based retail store that was founded in 1983 and is currently owned by Walmart. Sam's Club offers high-quality commodities at affordable prices (Our Company | Sam's Club).

Technology	Internet of things (IoT)/computer vision and deep learning
Description	Shoppers hate waiting in line. Sam's Club found a solution to this issue by developing a self-checkout application. This application allows customers to enter the retail store, scan a code on their mobile app, and pick up their items and leave. The app is equipped with sensors and computer vision technology capable of tracking the items and instantly sending the customer the bills (10 Technologies That Can Change Retail Forever I CMO)
Stage	In market
Beneficiaries	Consumers	• Fast, secure, and easy payment methods • Eliminated queuing • Minimized human contact
	Employees	• Time-efficient • Labor-effective

Telecommunications

Abstract The telecom sector worldwide has continuously evolved and relied on innovation as a telephone provider from the days of operators and payphones to the industry that created much of the global internet infrastructure. Many people think that the telecom industry consists of products and services related to cell phones, internet connections, and cell towers. However, there are many different segments of this industry that work in the background to make sure that when someone calls their family in a different country or search for the number of their favorite restaurant, this experience goes on with minimal friction.

Keywords Eco-friendly · Biometric scanning · Remote car control · Online car tracking · Precision market insights · Virtual assistant · Chatbot · Standardized service · Movement tracking · Material optimization · Touchpads · VR content · Self-optimizing networks · Predictive analytics · Predictive maintenance · Interactive voice response · Service optimization · Real-time threat detection · Internet connected devices

The telecom sector worldwide has continuously evolved and relied on innovation as a telephone provider from the days of operators and payphones to the industry that created much of the global internet infrastructure. Many people think that the telecom industry consists of products and services related to cell phones, internet connections, and cell towers. However, there are many different segments of this industry that work in the background to make sure that when someone calls their family in a different country or search for the number of their favorite restaurant, this experience goes on with minimal friction.

While the telecom industry has seen and continues to see tremendous growth, it faces many challenges. One of the main problems of the telecommunication industry is the security of the networks, and this has become a significant priority for the service providers. As new technologies are advancing, new threats are emerging that pose many challenges that require several operational and technical innovations to meet network customers' safety expectations. With the abundance of emerging technologies, various quality services are developed in telecom companies and internet service providers (ISP)s, which decrease the profit margins and blur the

© Springer Nature Switzerland AG 2023
M. Anis et al., *Mapping Innovation*,
https://doi.org/10.1007/978-3-030-93627-3_13

lines between telecommunication companies and technology vendors. Therefore, companies should digitally transform their organization by examining the levels of ICT, information, and communication technologies and developing robust cross-functional interfaces with organizational flexibility. Due to the significant capital expenditure involved in the industry, you need a large customer base and, in most cases, considerable market share to be able to operate profitably. Also, telecom providers should regularly update the information technology and connectivity infrastructure so that they can focus on providing data and voice services at high quality and also be affordable and reliable at the same time (Top 5 challenges & trends in the telecommunication industry in 2020 | Racknap 2020).

One of the promising trends in the telecom industry is 5G networks, which will spark a faster waver of the Internet, but the technology is not yet fully developed. Moreover, telecom is deeply connected to artificial intelligence (AI) and the internet of things (IoT), allowing telecom devices to perform highly sophisticated functions. It can open up new streams of revenues for the industry (Top 5 challenges & trends in the telecommunication industry in 2020 | Racknap 2020).

The COVID-19 pandemic has a significant negative impact on many industries, such as retail and construction industries, while demonstrating the critical importance the telecom industry plays in our lives by keeping governments, businesses, and societies connected. People worldwide rely on telecom technologies to communicate with others, work from their homes, and attend educational institutions online (The impact of COVID-19 on the Global Telecommunications Industry | International Finance Corporation).

Value Chain

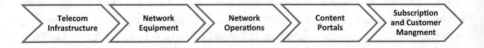

The telecommunications industry's value chain can be categorized into telecom infrastructure, software, network operations, distribution, and customer management.

Telecom infrastructure is comprised of manufacturers and providers of telecommunication equipment and tools that are used to create telecommunication services. Network equipment encompasses the development and software used. Network operations refer to the internal processes carried out to create and deliver offered services. The content phase refers primarily to distribution where telecommunication services are offered, online and offline, allowing consumers to search for, purchase, and even refund services. The final phase encompasses any aftersales customer services and database management and generally maintaining communications with customers.

Case Studies

Case Study 1

ICT and Renewable Energy

| Telecom Infrastructure | Network Equipment | Network Operations | Content Portals | Subscription and Customer Management |

- Country of implementation: Langfang, China
- Company: China Telecom

Founded in 1995, China Telecommunications Corporation also known as China Telecom is a state-owned company and is the largest fixed-line service provider in China and the third largest mobile telecommunications provider in China. They have over 216 million fixed-line subscribers and 43 million mobile subscribers.

Technology	Green technologies/wind/solar energy production	
Description	There is a part of all network infrastructures and telecom operators called base stations which is a fixed point of communication for cellular phones. Base stations account for between 50% and 80% of the energy costs that telecommunications firms incur. China Telecom is planning to use solar panels and wind turbines instead of the traditional nonrenewable sources such as gas and oil. Using the sun and the wind to generate the energy needed for these antennas to operate shall reduce the cost of the electric power that is consumed inside the base stations. Moreover, this will help clean the environment of all kinds of pollution associated with gas and oil; hence, it is an eco-friendly solution (Farah A., Naeem M., et al. 2016)	
Stage	Launching phase	
Beneficiaries	China Telecom	• Reduced energy costs • Avoidance of future regulatory pressure
	Customers	• Reduced telecom fees
	Chinese citizens	• Reduced pollution

Case Study 2

The State of Biometrics Solutions

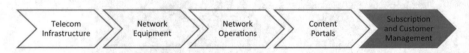

- Country of implementation: Nigeria
- Company: Airtel

It is the second largest telecom provider in Nigeria. Founded in 2001, it holds approximately a 30% market share with approximately 46.8 million mobile phone subscribers. It is a subsidiary company of Bharti Airtel who operates in 14 countries throughout Africa. Airtel became publicly listed on July 9th, 2019 with shares trading on the Nigerian Stock Exchange.

Technology	Biometrics/fingerprint scanning technology	
Description	Airtel had been facing a lot of criminals who use their SIM cards in blackmailing and kidnapping crimes. Therefore, Airtel uses biometric scanning when a user purchases a new SIM card; they developed a system that reads the user's fingerprint at the retail locations which sell these SIM cards so that people's identities are linked with all purchased SIM cards. Afterward, any SIM card that will be involved in any illegal activities will be identified with its user which shall leverage the security of the customers in Nigeria (Aware 2020)	
Stage	In market	
Beneficiaries	Consumers	• Improved safety • Decreased credit card fraud

Case Study 3

Cloud and Smart Home Service Platforms

- Country of implementation: Spain
- Company: Telefonica

Telefonica is a multinational telecommunication company that is based in Madrid, Spain. It provides mobile phone, broadband, fixed-line, and television subscription services in Europe and Latin America. It is a publicly traded company that was formerly owned by the Spanish government. It is one of the largest companies in the world, and it operates under several other brand names such as O2 in the UK, China Unicom in China, and Vivo in Brazil.

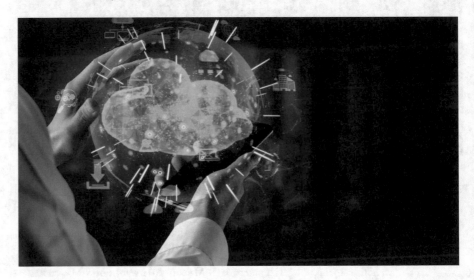

Technology	Platforms
Description	Telefonica launched a platform called Telefonica Aplicateca in 2009 to offer their customers in the small- and medium-sized enterprises space in Spain various entrepreneurship services. These services are provided by external vendors and they include cloud services, application solutions, service vendors, e-commerce services, marketing, customer relationship management, accounting, and office services. The way that Telefonica operates this platform is that it controls much of the pricing structure and products that vendors provide on the platform during the on-boarding to assure they meet their quality standards
Stage	In market
Beneficiaries	SME customers

Beneficiaries (SME customers):
- Increased access to services, e.g., white-label applications and services
- Supports high fee IT solutions

Case Study 4 orange™

Connected Cars Triumphantly Return to the European Home

Telecom Infrastructure ⟩ Network Equipment ⟩ **Network Operations** ⟩ Content Portals ⟩ Subscription and Customer Management

- Countries of implementation: Europe
- Company: Orange

Orange is a multinational French provider of telecom services that has over 266 million customers worldwide, and it is the 11th largest telecom provider worldwide. Its predecessors trace back to 1878 as the ministry of posts and telegraphs offered a telegraph network across France. It later evolved into France telecom in 1988 and was eventually publicly listed in 1997 with a follow on an offering of shares sold in 2000. The French government still owns about 25% of Orange.

Technology	Internet of things (IoT)/Internet of things (IoT) connectivity management	
Description	Ericsson, Orange, and Borgward, a historic German car brand revived by Chinese investors, partnered to bring the internet of things (IoT) to their new SUV. They leverage Ericsson equipment and Oranges European telecom network to provide internet services to Borgward's new cars. Borgward also uses Orange Business internet of things (IoT) services powered by Ericsson's internet of things (IoT) accelerator so that they provide a seamless implementation of internet connect services car such as remote car controls, online car quality tracking, in-car internet access, as well as other sophisticated connected car services (Ericsson 2019)	
Stage	Testing	
Beneficiaries	Drivers	• Remote car control • Engine start function • Online car tracking
	Passengers	• In-vehicle internet access

Case Study 5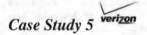

Big Data and Advanced Analytics in Telecom

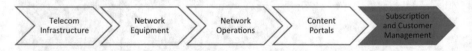

- Country of implementation: USA
- Company: Verizon

Verizon is the largest provider of mobile phone services in the USA as they offer fixed-line, cable television, broadband, and fiber optic internet services. Verizon Wireless was founded in 1999 as a joint venture between Vodafone and the Bell Atlantic company (provider of fixed-line services in the USA that later became Verizon communications). In 2014, Verizon wirelesses parent company Verizon Communications bought the remaining 45% stake from Vodafone 2014.

Technology	Big data and advanced analytics/clickstream analysis	
Description	To utilize the large amounts of data collected from their subscribers, Verizon changed their customer privacy policy so that they can anonymize and aggregate their customers' data which will be sold later to advertisers and firm companies. They launched a precision market insights business that anonymizes customer usage and location data and began selling it to sponsors, advertisers, media firms, and entertainment venues. It leverages Verizon's large subscriber base to provide a representative sample of the activities of the users with accurate trends in their behaviors (Huawei 2016)	
Stage	In market	
Beneficiaries	Marketing professionals	• Provides accurate, key data on online shopping habits • Supports targeted marketing campaigns

Case Study 6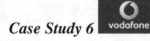

Artificial Intelligence (AI) in Telecom

- Country of implementation: UK
- Company: Vodafone

Vodafone Group PLC is a multinational provider of telecom services based in England. It operates networks in 25 markets and provides IT and consulting services in over 157 countries.

Technology	Artificial intelligence (AI)/machine learning	
Description	Telecommunication service providers receive tons of support requests for setup, installation, troubleshooting, and maintenance. Virtual assistants using artificial bots automate these support requests, which significantly reduce business expenses and increases customer satisfaction. In 2017, Vodafone launched its chatbot which provides its customers with an on-demand customer service experience. They can change phone plans, ask questions about their bill, and update their information or payment method and will be able to in the future do most of their interactions with Vodafone through Tobi the chatbox (Bernard M. 2019)	
Stage	In market	
Beneficiaries	Retail customers	• Increased customer satisfaction • Immediate access to Vodafone/support

Case Study 7

Blockchain
in Telecom

Blockchain in Telecom

Telecom Infrastructure	Network Equipment	Network Operations	Content Portals	Subscription and Customer Management

- Countries of implementation: Global
- Company: Bubbletone

Bubbletone is a startup that is trying to establish a link between developed market telecommunication companies and networks in other countries so customers can move around easily and have access to markets around the world at their local carriers' prices. It raised money to fund using a coin offering and is pursuing several projects that leverage blockchain technology.

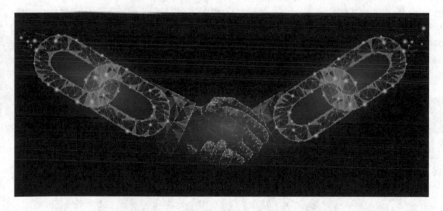

Technology	Blockchain
Description	To reduce the hassle of searching and subscribing to an aboard telecom service, Bubbletone created a marketplace using blockchain technology to allow customers to request wireless service abroad through their current provider at their regular prices for travel. Therefore, mobile providers will be offering international roaming and expand their user base without any additional costs (Vipin S., Sushant K., et al. 2020)
Stage	Launching phase

Beneficiaries	Consumers	• Reduced the hassle of finding telecom service abroad when traveling • Standardized services to avoid customer confusion • Reduced price uncertainty for travelers • Increased customer loyalty

Case Study 8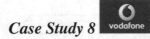

Vodafone UK

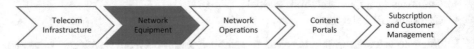

Telecom Infrastructure | Network Equipment | Network Operations | Content Portals | Subscription and Customer Management

- Country of implementation: UK
- Company: Vodafone

Vodafone Group PLC is a multinational provider of telecom services based in England. It operates networks in 25 markets and provides IT and consulting services in over 157 countries. It is the eighth largest company that is listed on the London Stock Exchange. It has over 313 million mobile phone subscribers. Vodafone evolved from the UK's largest producer of military radio components, Racal Strategic Radio, which was founded in 1982.

Technology	Internet of things (IoT)/heat mapping using cell data	
Description	Vodafone is using a cellphone on the Vodafone network as sensors to track movement data to produce heat maps of people's movements during the coronavirus crisis. Heat maps are data visualization method that illustrates the magnitude of data as a color in a two-dimensional plot. There, Vodafone is producing maps to show where people are gathering and improve further insights about their behaviors (Sharma 2017)	
Stage	Launching phase	
Beneficiaries	Governments	• Provision of key data • Improved safety

Case Study 9 orange™

3D Printing in Telecommunications

- Countries of implementation: Europe
- Company: Orange

Orange is a multinational French provider of telecom services that has over 266 million customers worldwide, and it is the 11th largest telecom provider worldwide. Its predecessors trace back to 1878 as the ministry of posts and telegraphs offered a telegraph network across France. It later evolved into France telecom in 1988 and was eventually publicly listed in 1997 with a follow on an offering of shares sold in 2000. The French government still owns about 25% of Orange.

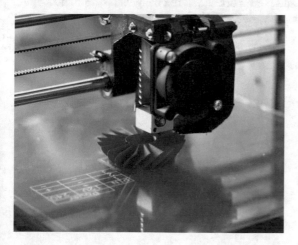

Technology	3D printing
Description	Eco Connection from Orange is using 3D printing to create parts that will reduce material costs and reduce material waste in low-volume production. For example, they are planning to use additive manufacturing to create wind turbines as a new source of power. Normal wind turbines are very expensive to mass produce. However, using 3D printed blades can improve efficiency and increase mobile connection performance and radically enhance their services (Plewa 2019)
Stage	Launching phase
Beneficiaries	Consumers	• Decreased material costs • Reduced material waste in low-volume production

Case Study 10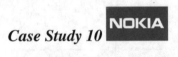

Nanotechnology in Telecom

- Country of implementation: UK
- Company: Nokia

Nokia is originally a Finnish telecommunication company focusing on producing network equipment, handsets, and other consumer electronics. It was founded in 1865 and now operates in over 100 countries with over 103,000 employees. It is publicly listed on the Helsinki and New York Stock Exchanges. It originally started as a pulp mill and diversified into a variety of different industries.

Technology	Nanomaterials/stretchable electronic skin
Description	Stretchable and compatible substrates can be exploited in developing elastic touchpads. These touchpads can stretch and conform to the human body and also enable new interfaces for users. This stretchable electronic skin forms a 50-nanometer depth of evaporated gold film to act as a conductor which is inserted in the rubber. This gold film will stretch with the skin of the user and return to its first shape (James M. 2015)
Stage	Testing phase
Beneficiaries	Nokia	• Improved potential revenue • Potential to sell product licensing and patents
	Cambridge University	• Potential to become a leading hub for education

Case Study 11 Discovery

Biometrics in Payment Security

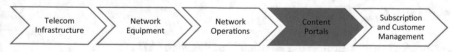

- Country of implementation: USA
- Company: Discovery Channel

The Discovery Channel is a television network founded in 1982. It began broadcasting in 1985 with an initial investment from the BBC, Allen Company, and Venture America. It was initially focused on purely educational programing and later evolved into producing more entertaining content. Discovery Channel was eventually acquired by Walt Disney.

Technology	Augmented reality (AR)/virtual reality (VR)	
Description	Given that much of Discovery Channel's nature content is already recorded on high-resolution/quality equipment, Discovery decided to leverage this content and be one of the first companies to offer virtual reality content through their Discovery VR application on mobile phones. Consumers can watch at their leisure as the content is already recorded and distributed through the application. Since the Discovery Channel is an educational one, this should be enhancing the space of education in virtual reality (Aaron 2016)	
Stage	In market	
Beneficiaries	Advertisers	• New advertising trends, potentially more compelling than normal television ads

Case Study 12   AriaNetworks

Network Optimization Using Artificial Intelligence (AI)

- Country of implementation: UK
- Company: Aria Networks

Aria Networks provides capacity management and network planning software. The incorporation offers a platform that provides network modeling, forecasting, routing, and optimizing communication networks and services for mobile, fixed-line, and cloud service providers.

Technology	Artificial intelligence (AI)/network optimization	
Description	Certified security professional uses artificial intelligence to develop self-optimizing networks which allow the operators to automatically optimize networks' quality according to traffic information by region and time zone. Aria Networks uses AI algorithms to search for hidden patterns inside the data that shall spot and predict any possible network anomalies. Therefore, it allows the system to proactively repair the problems before the customer is negatively affected by them (Liad 2020)	
Stage	In market	
Beneficiaries	Network providers	• Eliminates need to regularly check network services • Faults are detected and fixed automatically • Reduced overhead maintenance costs
	Customers	• Improved service • Increased customer loyalty

Case Study 13 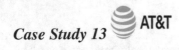 **AT&T**

AI in Predictive Maintenance

| Telecom Infrastructure | Network Equipment | Network Operations | Content Portals | Subscription and Customer Management |

- Country of implementation: USA
- Company: AT&T

AT&T is an American multinational incorporation based in Whitacre Tower in Texas. It is one of the largest telecom companies that provide mobile telephone and fixed telephone services.

Technology	Artificial intelligence (AI)/predictive maintenance
Description	Predictive analytics uses AI to help telecommunication companies provide broader and higher speed services by exploiting data and sophisticated algorithms so that they can foresee future results based on historical data. AT&T is using AI to support its maintenance procedures as the company is testing its drones to expand its LTE network coverage. They do so by analyzing the data captured by drone videos which helps them in the maintenance of their cell towers by anticipating failure based on patterns that can be proactively repaired automatically (Liad 2020)

Stage	In market	
Beneficiaries	LTE networks	• Expansion of LTE coverage
	Maintenance teams	• Data captured may predict possible failures

Case Study 14

Preventive Maintenance in Telecom

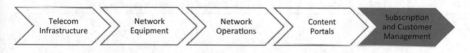

* Country of implementation: The Netherlands
* Company: KPN

It is a Dutch landline and mobile communication company that first started as a public telecom company.

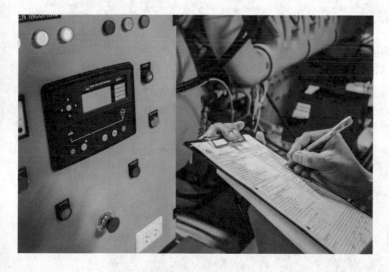

Technology	Artificial intelligence (AI)/preventive maintenance	
Description	Preventive maintenance is the regular maintenance of certain equipment to decrease the likelihood of failure taking place. Preventive maintenance is not only effective on the network side but also very effective on the customer's side. KPN studies the notes written by its contact center agents and analyzes them to get insights and do changes to its interactive voice response (IVR) systems. Furthermore, they also track the customer's at-home behavior, with their permission, like switching modem channels as this might accentuate a Wi-Fi issue (Liad 2020)	
Stage	In market	
Beneficiaries	Contact centers	• Improved voice response quality • Decreased need contact center personnel
	Maintenance team	• Potential to detect service fault

Case Study 15 / APACHE

Data-Driven Improvement in Services

- Country of implementation: USA
- Company: The Apache Software Foundation

ASF is an American company that was formed in 1999 to be an open-source community for developers. They produce their software under the terms of Apache License as free and open-source software.

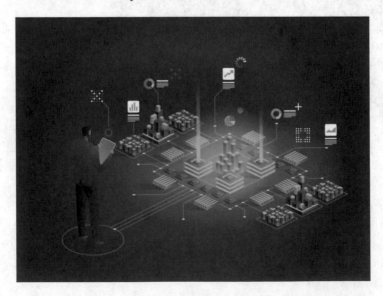

Technology	Big data and analytics
Description	Telecoms share data among different cell towers, users, and processing centers, and because of the large amount of data, it is vital to process it near the sources and then efficiently send it to the various data centers for further use. The Apache Software Foundation has published MapR Event Store which is a messaging system that is very efficient in transporting significant amounts of data and then sends it across to different geographical distribution data centers. Therefore, MapR Event Store works at a central data center to make events available in real time which means they can monitor the global performance and react accordingly to improve customer services. This helps them to understand how and where problems are affecting customers. It also helps them to optimize crow-based antenna by quickly monitoring network changing patter and making suitable reconfigurations (Carol 2017)
Stage	In market
Beneficiaries	Telecom companies	• Optimize services • Consistent monitoring • Preventive maintenance for dropped calls • Improved network coverage, bandwidth issues, service wait time, switching, and capacity use

Case Study 16 ⦿ NTT Group

Threat Detection

- Country of implementation: Japan
- Company: NTT

Nippon Telegraph and Telephone Corporation is a Japanese telecom company that is based in Tokyo and ranks the fourth largest telecom company worldwide in terms of its revenues.

Technology	Big data and analytics
Description	NTT Group is a pioneer in security services as they provide threat intelligence services to their clients. Their platform collects and analyzes significant amounts of data from firewalls, logs, and network devices. As the data volume increased, it became harder for them to continue with their existing database solution and they needed to improve their scalability. They developed a new database solution to achieve the required scalability and gather the sheer volume of data to be analyzed and used in predictive analytics to detect real-time threats (Carol 2017)
Stage	In market
Beneficiaries	Security professionals

Beneficiaries benefits:
- Decreased complexity
- Fewer moving parts
- Multiple clusters to be streamed in one cluster
- Improved reliability of data replication

Sprint ⟩⟩
Now part of
Case Study 17   **T**·Mobile·

Sprint's Dedicated Internet of Things (IoT) Network

- Country of implementation: USA
- Company: Sprint

Sprint Corporation is an American telecom company for wireless and internet services. It is owned now by T-Mobile and is ranked the fourth largest mobile network operator in the USA.

Technology	Internet of things (IoT)	
Description	Sprint has created a whole independent internet of things (IoT) network that is called Curiosity internet of things (IoT). It is a virtualized and distributed core network that is 5G-ready. It allows for a different organization the ability to manage all Internet-connected devices together. It also speeds up the development of internet of things (IoT) devices and time to get immediate analytics and intelligence on generated data. Telecom uses this Curiosity internet of things (IoT) in the visualization of vehicle fleets and tracking products in transit. SDxCentral partnered with Sprint and used Curiosity internet of things (IoT) technology and they reported that they were able to track valued assets from where they are shipped from the manufacturer to their store location and know what is lost and what is a defect (Connor 2020)	
Stage	In market	
Beneficiaries	Inventory managers	• Device management • Product tracking

Textiles

Abstract The textile industry is one of the oldest and largest industries in the world and has been forecasted to have a global population of 8.1 billion by 2025 (Fiber2Fashion, Textile Sector Gets Future Ready, 2014). As generations evolve and change, their needs also change. Accordingly, innovations in the textile industry become a necessary adaptation, and so overcoming production challenges and seizing new opportunities within a globalized economy have become a priority.

Keywords Design customization · Fabric pattern recognition · Top-performing content · Carbon emission reduction · Eco-friendly material · Chemical resistance · Drying processes · Energy classification system · Predictive maintenance · Global tracking · Digital inventory management · Time optimization · Performance tracking · Automated sewing machine · Smart garments · Robot-designed garments · Antimicrobial fiber · Reduced discoloration · Controlled moisture

The textile industry is one of the oldest and largest industries in the world and has been forecasted to have a global population of 8.1 billion by 2025 (Fiber2Fashion 2014). As generations evolve and change, their needs also change. Accordingly, innovations in the textile industry become a necessary adaptation, and so overcoming production challenges and seizing new opportunities within a globalized economy have become a priority.

In 2001, China joined the World Trade Organization (WTO). By 2004, China had gained access to markets that were previously difficult to penetrate. With the elimination of the quota system, a system that limits how many immigrants can enter the USA, they posed a considerable threat toward global manufacturers in the industry. China has mass productivity that provides its textile industry with economies of scale unmatched by its rivals. It has flooded global markets with textile products at unbeatable prices. This trend became a red flag to other manufacturers signaling a need to differentiate their products.

Additionally, concern for the environment has been a growing trend, and adopting environmentally friendly practices has become one of the more significant challenges this industry is trying to overcome. Textile manufacturing has been perceived

© Springer Nature Switzerland AG 2023
M. Anis et al., *Mapping Innovation*,
https://doi.org/10.1007/978-3-030-93627-3_14

as an environmentally hazardous industry due to its dependency on chemical treatments.

As consumers become more environmentally conscious, it has become necessary for manufacturers to integrate sustainable approaches within the sector.

R&D-based innovation has become a necessity for textile companies to allow them a competitive advantage in the current globalized economy and has resulted in infusing the industry with great opportunities. The resulting plethora of advancements expanded the industry's horizon to encompass differentiated products to satisfy the market's ever-changing needs. Technological disruption has impacted the industry's supply chain and renovated the synergy between all parties, starting from the supplier to the end consumer.

Digital technology eased communication across the value chain, from asset tracking and inventory management upstream to artificial intelligence (AI) software that handles marketing and sales operations downstream. Technology affected industry demand trends as new products generated new kinds of demand. For example, smart textiles and camouflage fabrics started to emerge creating new markets that in turn created new segments that started to generate demand for such products. Trends such as recycling create green technology opportunities. Running a facility at zero waste helps companies benefit from cost reduction in addition to increasing their brand equity (Fiber2Fasion 2014).

Furthermore, energy consumption has been another trend of interest. The noticeable depletion of natural resources sparked the need to utilize renewable resources for energy as producers within the industry seek ways to become more efficient in their energy usage. Not only does this lower cost but it also promotes sustainable corporate social responsibility. Renewable energy sources have become lucrative manufacturing alternatives, thanks to technological advancements that provided cleaner and cheaper options than fuel-based electricity. Textile producers worldwide are adapting these technologies into their manufacturing, as they have become more efficient, reputable, and credible. These technology trends, associated with what has been dubbed the fourth industrial revolution, will have an impeccable impact on how the industry will perform.

As of April 2020, the International Monetary Fund (IMF) forecasted a 3% contraction in the worldwide economy (FAO 2020). An unanticipated recession will trigger a reduction in the gross domestic product (GDP) across all countries between 2 and 10%. Such an impact will add to the challenges the textile industry will face. Unanticipated changes in consumer behavior and emerging needs will inevitably take their toll on the industry. Innovation and research will need to be deployed to help industries cope with the difference in their external economic environment.

On the other hand, opportunities have risen for textile industries, including the increase in demand for masks, patient gowns, towels, bed linens, etc. Also, businesses are now designing and developing new technologies to protect clothes against the virus (Internal Labor Organization).

Value Chain

The textiles' industry value chain involves five main phases: product production of inputs, manufacturing, marketing and display, distribution and sales, and waste management.

The first phase is concerned with the processing of raw materials into inputs, suitable for the industry such as threads, cloths, buttons, and dyes. These inputs are then processed further by manufacturers, during the second phase, into their final form to be used by the consumer. Products are then shipped to retail stores to be promoted and displayed (marketing and display phase) and then sold and/or shipped to consumers (distribution and sales phase). Finally, the waste produced, by the industry, is either discarded/recycled at the final waste management phase.

Case Studies

Case Study 1 ⌒ HER MIN

Artificial Intelligence (AI) for Apparel Design

- Country of implementation: Taiwan
- Company: HerMin Textile

Mr. Chen Wen-Jin founded HerMin Textile Co. LTD in 1976. In the early transformation stages, HerMin's Tainan focused its production on "woven plaid fabric," using all-natural fibers – "cotton" based mostly, followed by silk, linen, and wool, among others. HerMin Textile developed fabrics using state-of-the-art fibers, such as Lycra, Nylon, Organic Cotton, recycled polyester, Sorona, polypropylene, Xylitol and UMORFIL®, and more. These new fabrics have been recognized by consumers around the globe for their multifunctional features in comfort, refinement, and style (HerMin 2020).

Technology	Artificial intelligence (AI) and machine learning/cloud	
Description	HerMin textile provides labeled images of textile designs to a CloudMile data engineer who uses Google Cloud Data lab to undertake data analysis and develop machine learning models. With this system, rather than locating designs and sending individual digital photographs to customers, a process that may take several days if multiple images or designs are involved, HerMin Textile can share several patterns at once. HerMin Textile wanted the machine learning project to power a mobile application that would modernize communications among designers. The result is the ability to meet customers' requirements at an increased capacity, as the machine learning system will enable the business to select similar or relevant patterns for their use (Google Cloud 2020)	
Stage	Idea	
Beneficiaries	Fabric manufacturing	• Faster coping against fast fashion businesses and new fashion designers • Establishing a system to enable businesses to create more than 10,000 designs in the next 2 years • Automated warehouse sample selection • Enhanced designing speed up to 25% • Machine learning that improves the process of matching client selections
	Designers	• Enhanced designer communication through the app's use • Faster new design generation
	Customers	• The faster design selection process • Wider variety of design selection

Case Study 2 COGNEX

Artificial Intelligence (AI) for Inspection and Quality Control

- Country of implementation: USA
- Company: Cognex

Cognex is a leading provider of vision sensors, systems, software, and inspection systems that enhance innovation. It was founded by Dr. Robert J. Shillman in 1981. They aim to help other businesses improve their product quality, lower costs, and exceed customer expectations by guaranteeing a higher quality product.

Technology	Artificial intelligence (AI)/automation and machine vision
Description	Cognex ViDi is designed using artificial intelligence and machine vision to enhance fabric pattern recognition. This technology can easily track and observe fabric patterns, including weaving, knitting, finishing, braiding, and printing. Initial programming involves scanning defined images of fabric patterns to identify high-quality fabrics and their distinguishing features and to create general reference models including yarn properties, patterns, colors, possible imperfections, printing quality, and any irregularities, such as ink stains. Within a 2-week analysis period, machines can recognize any defects or bad samples which saves workers from the arduous task of assessing hundreds of yards of material manually (Bharadwaj 2019)
Stage	In market

Beneficiaries	Manufacturers	• Elimination of production errors • Increased quality control efficiency • Automated and scales complex inspection application • Easy-to-use in-sight software platform • Faster and more reliable process • Higher customer satisfaction and retention • Lower manufacturing costs • Quality product • Time-effective
	Customers	• Fewer product returns • Standards achieved at the first trial • Consistent quality

Case Study 3

Artificial Intelligence (AI) for Digital Marketing Creative

| Raw Material/ Inputs (Thread & Fabric) | Intermediary Product Production (e.g., Buttons, zippers) & Final Product Production | Marketing & Display | Distribution & Sales | Waste Management |

- Country of implementation: USA
- Company: Persado

Persado is a Marketing Language Cloud company that uses artificial intelligence language. Persado's Marketing Language helps other brands and businesses increase purchases and use while at the same time maintaining long-term consumer relationships. It works in several industries including advertising, data analytics, market research, brand marketing, and many others.

Technology	Artificial intelligence (AI) and machine learning/data analytics
Description	By combining words with data, Persado breaks down marketing creative into six key elements and then runs experiments on thousands of potential message combinations to generate the best-performing content to speak to each customer across their entire journey. This provides customers with the confidence that their digital marketing is fully optimized for brand engagement and revenue performance (Persado)
Stage	In market

Beneficiaries	Apparel manufacturer	• Delivers effective brand messages • A system that learns and improves continuously • Measures the impact of certain phrases • Analyzes millions of messages and draws insights through artificial intelligence (AI) and deep learning • Communicates with customers across digital platforms and delivers customized messages • Increases brand loyalty and improves brand perception
	Customers	• Identifies the right creative message that delivers the brand's message increasing the possibility of conversion • Updates and notifies different digital media platforms

Case Study 4

Green Technology Energy Consumption

- Country of implementation: USA
- Company: Arvind

Arvind Co. is a fashion powerhouse that designs and builds new age homes. It is also a global leader in garment manufacturing that is transforming water management within the industry. It is also a pioneer in denim manufacturing that explores and utilizes advanced materials. Moreover, it is a textile technology manufacturer that is also delivering state-of-the-art engineering solutions. This company has the continuous potential to change lives and make a difference globally (Arvind).

Technology	Green technologies/recyclable
Description	The Energy Conservation Cell was created to ensure effective implementation of the "Power of Nature Policy" supervised by the CEOs of each business. The efforts of controlling energy conservation range from simple switching to LED lighting across plants, utilizing natural light by building transparent roofs, and other different methods, such as adding heat recovery systems (Arvind) Arvind plans to reduce carbon emissions by 30%, with the installation of rooftop solar projects across its facilities in three cities, by shifting from coal to renewable biomass for boilers. As the company ramps up the three-phase installation, it plans to target 40 MW captive solar capacity. Rooftop solar installations across its facilities will generate 22 million units (KWh) of power per annum. It will contribute to the reduction of 20,000 tons of carbon emissions annually as well as over 5,00,000 tons of carbon emissions over its lifetime. Once the capacity of 40 MW is reached, overall generation will exceed 55 million units per year and will reduce carbon emissions by 50,000 tons per annum (PTI 2019)
Stage	In market
Beneficiaries	Manufacturer

	Manufacturer	• 60,400 k cumulative saving through daylight harvesting • Over 3 million kWh saving through energy-saving LED lamps • A decrease of 11.78% in gas emissions • Saving 1.4 Mn kWh/annum with energy-efficient ones • Solar panel's overall generation will exceed 55 million units per year • Reduce carbon emissions by 50,000 tons per annum. • More cost-efficient through energy savings • A recognized brand for sustainability initiatives
	Utility	• Decrease load on the power grid • Sell excess power to the power grid • May benefit subsidies on the utility electricity that is being removed "Egypt"
	Client	• Potential cost savings

Case Study 5

Green Technology Waste to Zero

- Country of implementation: USA
- Company: A&E

  A&E works as part of Elevate Textiles, which is a portfolio company of Platinum Equity that has worldwide operations in several countries including Mexico and China. Elevate Textiles helps global textile brands, such as Cone Denim, Burlington, Gütermann, and Safety Components. A&E provides its customers with high-quality goods by following modern quality concepts and practices (Amefrid).

Technology	Green technologies/recyclable
Description	Core® ECO100 consists of 100% recycled polyester staple wrap and continuous strand recycled core. It provides great aspects of a sustainable sewing thread without abandoning sewing performance or thread color. Wildcat® ECO100 consists of 100% recycled and textured polyester and provides good elasticity, softness, and strength creating better-performing and longer-lasting, softer seams while delivering an eco-friendly product. The final material is Anefil Poly® ECO100 which provides a 100% recycled polyester sewing strand in a twisted multifilament establishment and provides good chemical resistance. It allows for the adaptability of producing an environment-friendly sewing thread into a diverse mixture of non-garment applications (Amefrid)
Stage	In market

Beneficiaries	Yarn manufacturer	• Enhance customer loyalty • Meet client requirements for sustainable products • Gain competitive advantage for being an environmentally responsible company • Reduce waste production • Increase brand awareness and exposure
	Environment	• Reduce the impact of plastic waste • Reduce the consumption of new materials
	Customers "Apparel manufacturers"	• Promote their products using the yarn manufacturer brand's green technology • Better brand perception for being an environmentally responsible product

Case Study 6 Jeanologia
The Science of Finishing

Green Sustainability Measuring Software

- Country of Implementation: Spain
- Company: Jeanologia
  Jeanologia is a company that provides textile solutions by designing technologies that improve and increase productivity, reduce water and energy consumption, and eliminate damaging emissions and waste, thereby guaranteeing zero contamination. It was founded by Jose Vidal and his nephew Enrique Silla in 1994 (Jeanologia).

Technology	Green technology/software platform
Description	EIM (Environmental Impact Measurement) software is the only software developed in the market that measures environmental impact in the textile industry. Energy consumption computes mechanical energy that is being used to run the machines and caloric energy to heat the water that is used in the washing processes and the air that is being used for drying processes. Environmental Impact Measurement takes into consideration both energies and brings them to a one-unit measure. The total energy consumption of one finishing process will be the sum of the energy consumed in every step. Another aspect that is taken into consideration is the environmental hazards caused by chemicals used. The chemical products are described and ranked as chemicals of high, medium, or low impact on the environment which follows the self-classification system produced by TEGEWA. The total chemical impact of one process will be the sum of all the chemicals involved in it. Two important factors can affect the workers' health in the laundry; the first one is handling the chemicals and the second one is the operations they work on. To evaluate the chemical impact on workers' health, EIM takes into consideration the toxicology of every product (Jeanologia)
Stage	In market

Beneficiaries	Apparel manufacturers	• Sustainable goals for water and energy consumption • Safer work environment • Measured and tracked sustainable goals • Increased corporate citizenship behavior • Brand positioning • Cost-effective • Efficient water and energy consumption • Environmentally friendly
	Workforce	• Safer working environment
	Customers	• Awareness of brand sustainability

Case Study 7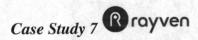

Internet of Things (IoT) Predictive Maintenance

- Country of implementation: Australia
- Company: Rayven Internet of Things (IoT)
  Rayven digitally transforms industrial companies and enables them to transform their data into desired business outcomes. The internet of things (IoT) platform technology merges a lot of applications into one system, including device management, real-time data streaming, machine learning, and data visualizations. Rayven provides nontechnical users with a design interface using highly customized internet of things (IoT) solutions affordably and efficiently. Users can then ingest, combine, transform, and display data for a wide array of purposes. This company uses internet of things (IoT) and artificial intelligence (AI) solution development, from device hardware connectivity to security to data science (Rayven).

Technology	Internet of things (IoT)/cloud
Description	The technology designed is predictive maintenance (PdM) which is an important process for any industry aiming to predict equipment failure. Accordingly, industries can prevent potential downtime and major financial downfall. Constant monitoring for future failure enables businesses to be one step ahead and to plan maintenance before any failure occurs. By doing so it eliminates major downtime of assets or processes. Predictive maintenance (PdM) also reduces unplanned reactive maintenance and reduces the cost associated with preventive maintenance (Rayven)
Stage	In market
Beneficiaries	Fabric/yarn manufacturers

Beneficiaries benefits:
- Reduced time equipment maintenance time
- Reduced production hours lost to maintenance
- Reduced unplanned stoppages
- Increased predictability
- Improved uptime by 9%
- Reduced costs by 12%
- Reduced safety and environmental risk by 14%
- Extended asset lifetime by 20% (Christiansen 2019)

Case Study 8 **Actility**

Internet of Things (IoT) Asset Tracking

| Raw Material/ Inputs (Thread & Fabric) | Intermediary Product Production (e.g., Buttons, zippers) & Final Product Production | Marketing & Display | Distribution & Sales | Waste Management |

- Country of implementation: France
- Company: Actility is a company that aims to employ the internet of things (IoT) to service cities, citizens, industries, and communities all over the world. This company transforms activities, businesses, and processes by guaranteeing structured and secured data transfer, sensor management, operational support system, management of data flows, and monetization. Actility enhances customer ecosystems by supporting developers and device makers preparing their low-power wide-area product for the market and providing an e-commerce marketplace offering global distribution to solution providers (Actility).

Technology	Internet of things (IoT)/digital software, GeoCountry of implementation
Description	ThingPark internet of things (IoT) connectivity platform enables integrated and scalable internet of things (IoT) network infrastructure managing both private and public LoRaWAN™. The activity offers a reliable and flexible low-power internet of things (IoT) country of implementation platform which enables global tracking. Small, battery-powered tracking devices with embedded LoRaWAN™ are securely positioned onto containers, delivery cars, trucks, and many more. These small tracking devices collect real-time data about their country of implementation. They run for years at a time without a required battery change and can work both indoors and outdoors. The collected data is then securely sent to Actility's internet of things (IoT) connectivity platform which is connected to a user's application or a company's existing logistics information system. Using this information users can see accurate information about the exact country of implementation of a property and its estimated speed and arrival time based on current travel conditions. It also sends alerts in case of unexpected events, for example, if a container traveling by boat or rail is delayed (Actility)
Stage	In market

Beneficiaries	Distributors	• Supply chain visibility • Streamlining field operations • Reducing delay costs • GPS-free country of implementation monitoring
	Customers	• Improved product tracking • Increased customer satisfaction • More efficient delivery

Case Study 9

Internet of Things (IoT) for Inventory Management

- Country of implementation: USA
- Company: GEP SMART
  GEP SMART is a leading company in technology innovations; it is also a pioneer in cloud-based acquisition and supply chain solutions for blue-chip enterprises globally. GEP produces and designs intelligent software to help supply chain organizations at market-leading enterprises across the world to enhance productivity and performance and turnaround times and increase savings. GEP is one of the first developers to realize the potential of cloud, touch, and mobile technologies. Furthermore, it tailors solutions for businesses that include engaging, intuitive experience applications (GEP SMART).

Technology	Internet of things (IoT)/digital software
Description	This technology was designed with a digital inventory management system to manage accounts across the supply chain in real time, observe stock levels and supplier performance, send alerts for restocking, create smart order restoration based on recommended inventory levels, and predict demand. This helps enterprises optimize their inventory and minimize costs. GEP SMART makes use of sophisticated artificial intelligence to combine the complete range of inventory management activities, such as purchasing, payments and shipment tracking, and more, into a single platform (GEP SMART)
Stage	In market

Beneficiaries	Manufacturers	• Improved order accuracy and ETAs, minimizing "out-of-stock" situations • Ability to analyze stock demand • Audit inventory with cycle counts, periodically/yearly • Analyze reorder point and quantity • Provide in-context analytics and insights
	Suppliers	• Efficiently plan production and shipping based on inventory level • Higher efficiency of coordination streamlines supply chain operations

Case Study 10

A Platform for Sales Management

- Country of implementation: Italy
- Company: Datatex

  Datatex is a leading supplier for the textile industry of IT software solutions. With one of the largest install bases of apparel software in the world, it has customers in 45 countries and 5 continents. Datatex's goal is to enhance other businesses by concentrating on textile and garment planning and arranging functions and supplying very comprehensive and precise costing functionality. This company's textile enterprise resource planning (ERP) platforms are the result of 35 years of experience working with the newest information technologies. Datatex's accurate planning capabilities are a key success factor for its customers (Datatex).

Technology	Platforms/digital software
Description	Datatex has developed technological solutions that maximize production. The first step is pricing competitively, followed by customer support management. An important demand for Datatex is the capability to confirm both availability in the warehouse and the supplier's processing and shipping times to ensure delivery dates. To further ensure good sales process management, businesses are encouraged to problem solve and follow the product's path carefully. This provides the customer details about the stage and localization of his/her order. During this process, Datatex manages returns, claims, and credit notes (Datatex)
Stage	In market
Beneficiaries	Textile manufacturers

Beneficiaries		
Textile manufacturers	• Management of customer quotations • Efficient product ordering • Verified product availability • Defined delivery times • Customer order status
Customers	• Accurate delivery dates • Real-time updates on products availability • Smoother return and claims procedures • Competitive pricing

Case Study 11 ❊ FOURSOURCE

Digital Platform for Marketing

- Countries of implementation: Germany
- Company: Foursource

  Foursource's goal is to solve the missing transparency of the highly shattered global sourcing markets and to make textile sourcing quicker, safer, and less costly. This company supplies marketers with solutions to considerably increase their cycle time and to purchase their value proposition to their customers. Foursource also follows a worldwide proposition with direct and sales partners in a large number of countries. Foursource is located in Berlin, Germany, where its technology and development center is located in Porto, Portugal (Foursource).

Technology	Digital platforms/big data
Description	Foursource offers free sign-up for both the buyers and manufacturers alike. The buyers can have full benefits from the service free of charge, and manufacturers enjoy limited benefits free of charge and an option to enjoy advanced benefits for a premium. Foursource database provides access to buyers from 120 countries with a retail value of 50 bn USD and access to more than 40,000 apparel manufacturers across 116 countries. Foursource has a strong validation process, through collaborating with accreditation bodies, to ensure manufacturer's compliance, to production efficiency, to encourage sustainable practices (Foursource)
Stage	In market

Beneficiaries	Apparel retail stores	• Access to the world's largest network of verified garment manufacturers • Thousands of detailed company profiles • Fast, structured, and comparable offers from relevant factories • Explore 1000 s of virtual supplier showrooms, wherever you are • Direct access to manufacturers • Source suppliers • Manage social compliance standards through Foursource certificate validations with OEKO-TEX, GOTS, WRAP, and Textile Exchange • Build and manage personal supplier network • Store data in one place • Confidentiality
	Apparel manufacturers	• 365 days a year 24/7 visibility • MOQ or certificates • Creation of professional company websites • Prominent Google ranking increases visibility • Direct consumer messaging • Relevant buyer RFQs • Direct buyer RFQs using structured forms

GARMENT IO

Case Study 12

A Platform for Production Management

• Country of implementation: Egypt
• Company: Garment IO

Garment IO supplies textile industry managers and owners with a plug and plays garment manufacturing management solution that combines software and hardware to give them a complete run-through of manufacturing processes, employee performances, and manufacturing cost calculation, in addition to the transaction detail and real-time data (Garment IO).

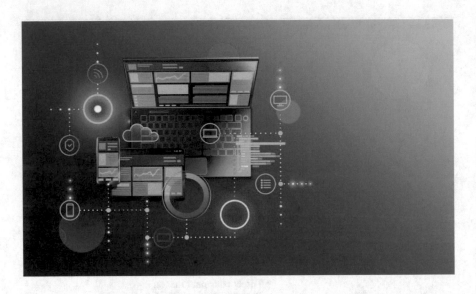

Technology	Platforms/big data
Description	Garment IO provides a complete data ecosystem for the client's industry. Using a cloud-connected proprietary device at every transaction point and modular cloud system, industries will be able to log, examine, and record every activity based on preference to defend strategic decisions (Garment IO)
Stage	In market
Beneficiaries	Garment manufacturers

Beneficiaries	Garment manufacturers	• Boosted production by 20% • Enhanced worker performance • Individual quality reporting • Real-time tracking through cloud technology • Maintenance module • Order tracking • Secured and encrypted data • Projected delivery dates
	Employees	• Tracked performance • Targeted improvement areas • A better engaged and motivational environment
	Garment retailers	• Tracked orders • More accurate delivery times • Improved quality

Case Study 13 WEARABLE X

Nadi X Smart Yoga Pants

- Countries of implementation: Available in 36 countries
- Company: Wearable X

  Wearable X is a fashion company founded by Ben Moir and Billie Whitehouse in 2013; it develops all its products using technology to provide a better quality of life. They have partnered up with many businesses to create different fashion tech products for different purposes, such as the alert shirt which is to allow fan engagement ad Fan Jersey X which was worn by thousands of fans during the Football season (Wearable X).

Technology	Biometrics/sensors, accelerometers, and Bluetooth low energy	
Description	This technology involves three components: the designed yoga pants with technologies embedded in them, a device called Pulse that is added behind the upper left knee, and an application; the pulse device is used to control the pulses in the pants and communicates with the phone by sending data about the person's movements to the application and then uses the algorithm of the yoga pose the person decided to use to identify it. Before the user begins with the yoga pose, he/she is asked to select the levels of intensity of the vibration to receive. The vibrations notify the user about the areas they need to improve to get the pose right and then the audio instructions will let them know if they have made it to the end of the pose; it is washable (Wearable X)	
Stage	In market	
Beneficiaries	Customers	• Customers notified of areas needing mending • New exciting technology • Increase fitness • Better health care • Audio-enabled • Ability to monitor and track fitness

Case Study 14

Lowry Sewbot

- Country of implementation: USA
- Company: SoftWear Automation
  SoftWear Automation is an innovative and autonomous company founded by Palaniswamy Rajan in 2007 and headquartered in Atlanta, Georgia; it designs machine vision, computing, and robotics technologies to enhance sewing for apparel, home goods, and footwear.

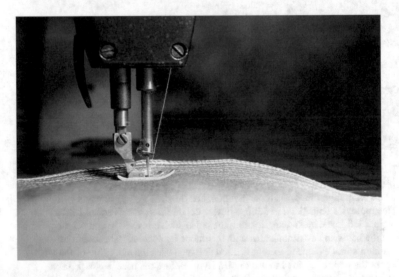

Technology	Artificial intelligence (AI)/robotics technology
Description	The Lowry Sewbot is a fully automated sewing machine that was designed and produced to sew goods. It can sew numerous types of products regardless of the size, material, or shape in different industries. It can manufacture pillows, towels, footwear, mattresses, bags, t-shirts, rugs, and vehicle mats. It cuts labor by 85% and can produce large quantities of products in just a few hours. For example, it can produce one pillow in under 75 seconds, one towel in under 60 seconds, and one t-shirt in under 22 seconds. It can easily adapt to any new product specification. One sewbot can produce the same number of t-shirts as 17 workers. The Lowry Sewbot can run 24 hours every day nonstop, and one operator can supervise six sewbots (SoftWear Automation)
Stage	In market

Beneficiaries	Industries	• Can produce different products for different industries • Can easily adapt to new instructions or new specifications to design a new product
	Stakeholders	• Increased efficiency • Increased production • Increased quantity and quality • 24/7/365 running ability • Multiple production lines • One sewbot can replace 17 workers • One operator can supervise up to six sewbots • Adaptation to various product specifications

Case Study 15 HEXOSKIN
HEALTH SENSORS & AI

Smart Clothing for Heart Monitoring

- Countries of implementation: Global
- Company: Carré Technologies Inc.

Carré Technologies Inc. or Hexoskin is a private company founded in 2006 by Pierre-Alexandre Fournier and Jean-François Roy; it designs smart clothes to measure and monitor health. It benefits multiple sectors, such as health care, pharmaceuticals, private organizations, and many more. It uses multiple technologies, including data science, artificial intelligence, sensors, and others (Hexoskin).

Technology	Biometrics/sensors, Bluetooth low energy, data analysis, accelerometers	
Description	Hexoskin Connected Health Platform is an end-to-end solution that provides sensors, software, and data science tools to facilitate monitoring for patients. Hexoskin designed a smart garment with sensors embedded in it to track and monitor cardiac, respiratory, sleep, and performance activities by connecting it to a phone or laptop. It tracks a person's heartbeat by producing an electro diagram which is a graph of voltage versus time to show the electric activity of the heart. It has its processor to send real-time data signal analysis to calculate statistics about heart rate, activity level, breathing rate, and volume, steps, and others. It is provided for men, women, and children and has 12–30+ hours of battery life; it is washable, very lightweight and breathable, and comfortable for the person wearing it (Hexoskin)	
Stage	In market	
Beneficiaries	Health-care sectors	• Provides an electro diagram with heart rate • Provides step count and breathing rate • Monitors respiratory rate and volume • Gives information about activity level • Available for men, women, and children
	Customers	• Lightweight garments • Comfortable • Washable • Easy to dry • Water-resistant • Breathable • Supports real-life everyday activities

Case Study 16

Water-Soluble Stiffening Fabric

- Country of implementation: USA
- Company: Sewbo, Inc.

  Sewbo, Inc. is an industry that offers automation for clothing manufacturers. It was founded by Jonathan Zornow in 2014 and headquarters is located in Seattle, Washington, USA. It aims to create and design high-quality clothes at lower costs.

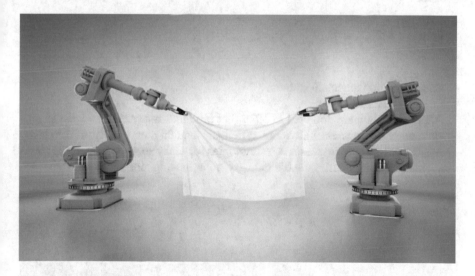

Technology	Artificial intelligence/robotics technology
Description	Sewbo is a robot that temporarily stiffens fabrics to allow off-the-shelf robots to design garments using rigid cloth like they are working with sheet metal. The fabric produced can be shaped and attached before sewing. The water-soluble stiffener, which was produced at the beginning, is removed at the end of the process by simply rinsing it in hot water to create soft clothing. The removed stiffener can then be reused. The average cost of plastic to stiffen one shirt is less than 10 cents; hence it is considered low cost (Sewbo 2016)
Stage	In market

Beneficiaries	Manufacturers	• Higher quality
		• Lower costs
		• Higher efficiency
		• Shortened the supply chain
		• Decrease delays
		• Increased output quantity
	Stakeholders	• Decreased workforce
		• Cost-effective
		• Time-efficient

Case Study 17 @te

Ecoaxis

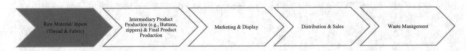

- Country of implementation: India
- Company: A.T.E. Group
  A.T.E. is a company founded in 1939 that produces many technologies to provide other businesses with more efficiency and higher quality of products. A.T.E. provides its services and technologies for many sectors, such as the printing and packing industry, textile industry, solar heating, and many more.

Technology	Internet of things (IoT)/data analytics
Description	Ecoaxis is an application that is designed using the internet of things that provides online monitoring and data analytics for producing fabric. It can monitor and collect information on the performance of many machines in a single country or many machines in multiple countries of implementation. Users are also able to enter details of the quality of the product which allows them to create a quality library for production planning. Moreover, they can easily track any malfunctions or failures in the machines and collect data on the production rate and speed. In other words, users can track the efficiency of the machines, follow production analysis, monitor performance, monitor energy consumption analytics, and many more. Users will be able to collect monthly and daily reports (A.T.E. Group)
Stage	In market

Beneficiaries	Industries	<ul><li>Efficiency tracking</li><li>Monitored rate of production</li><li>Quality control</li><li>Estimated average repair times</li><li>Monitored machine's performance</li><li>Monitored temperature and humidity</li><li>Measured energy consumption</li></ul>

Case Study 18

X-STATIC®

- Country of implementation: USA
- Company: Biomaterials
  Biomaterials is the leading company in bacterial management solutions and its headquarters is located in Great Lakes, Northeastern USA; it was founded by Joel Furey in 1997. It aims to design antimicrobial technologies to eliminate or decrease bacterial contamination.

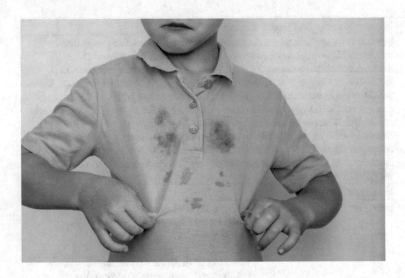

Technology	Biometrics/antimicrobial technology
Description	This technology is bonded with a layer of 99.9% pure metallic silver which in return produces an iconic shield that hinders the growth of bacteria and fungi on textiles. Hence, this reduces the risk of the patient being contaminated by bacteria. Soft surface textiles like the ones found at the hospital contain high bacterial contamination; therefore, fabrics made with X-STATIC® are designed with silver fibers which is an antimicrobial fiber and eliminates 99% of bacteria. It is designed to help hospitals ensure a bacterial management solution for soft surface textiles (Infection Control Today 2011)
Stage	In market
Beneficiaries	Patients

	Patients	• Eliminates 99% of bacterial and fungi elimination, therefore reduces the risk of being contaminated • Clothes smell fresher for longer • Clothes remain soft, flexible, and comfortable
	Hospital professionals	• Reduces contamination of hospital patients and staff as 78% of physicians' coats are contaminated and 60% of uniforms are contaminated (Noble Biomaterials) • Increases protection • Prevents infection

'TORAY'

Case Study 19 Innovation by Chemistry

&+™ – Fiber Made from Recycled PET Bottles

- Country of implementation: Japan
- Company: Toray Industries, Inc.

  Toray Industries, Inc. is a fibers and plastics company that combines nanotechnology into its solutions and makes use of different applications, including synthetic chemistry, internet of things (IoT) technology, and polymer chemistry. It was founded in 1926 in Tokyo, Japan. Toray Industries also promotes carbon fiber materials, medical products, IT-related products, and others.

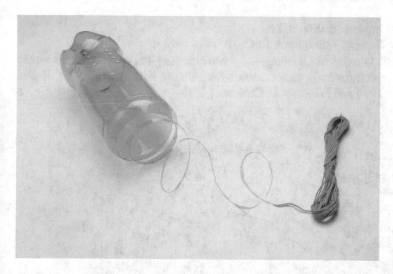

Technology	Green technologies/recyclable
Description	Toray Industries produced & + ™ which is a new fiber made from recycled PET bottles. These PET bottles are first collected throughout Japan then organized, cut, and cleaned in an alkaline solution. The produced material is then filtered and melted using a designed technology, which reduces impurities and discoloration of the recyclable fibers, to produce high-purity pellets which then go through intensive control inspections. Toray then uses its fiber manufacturing technology to transform the high-purity pellets into high-quality polyester fibers (Toray Industries, Inc)
Stage	In market

Beneficiaries	Manufacturers	• Environmentally friendly • High-quality products • Recyclable • Authenticity
	Environment	• Recycling • Creates a sustainable society

Case Study 20

Optimal Thermal Comfort

- Country of implementation: USA
- Company: Outlast Technologies LLC
  Outlast Technologies LLC is a privately held company that provides smart textile solutions to control body temperature by regulating heat and moisture. It was founded in 1990 and is located in Colorado, USA.

Technology	Biometrics/Thermocules	
Description	This textile is designed by printing the phase change materials (PCM) Thermocules onto the fabric where acrylic, polyester, or viscose fibers are melted with Thermocules PCM. Each type of fiber has its benefits and application, for example, acrylic fiber has high resistance against sunlight degradation and is used in socks or hats; viscose fiber provides softness and comfort and is used in shirts or dresses, and polyester fiber is a temperature-managing fiber. Fiber can absorb, store, and release excess heat. Therefore, when the body's microclimate begins to change, Thermocules can always regulate the skin's microclimate. For example, if the skin becomes hot, the heat is absorbed, and if it becomes cold, then heat is released to keep the body's temperature stable. This technology can control heat while at the same time controlling the production of moisture before it starts (Outlast 2018)	
Stage	In market	
Beneficiaries	Users	• Controlled and maintained body temperatures • Controlled moisture production

Tourism

Abstract Tourism is one of the world's largest industries and, for many nations, a driving economic force. Tourism injects money into the economy through hotel bookings and sales in local businesses. This industry has created numerous job opportunities in transport, hospitality, and construction as it encourages the upkeep of infrastructure and boosts economic growth alongside economic and cultural development (Tourism Industry: Everything you need to know about Tourism | REVFINE).

Keywords Customized experience · Search optimization · Cost-effective flights · Facial recognition · Navigation · Gamification · Baggage tracking · AI trip planning · Autonomous suitcase · Bluetooth-enabled locks · Humanoid robots · Speech recognition · Virtual touring · Speech-enabled devices · Trip personalization

Tourism is one of the world's largest industries and, for many nations, a driving economic force. Tourism injects money into the economy through hotel bookings and sales in local businesses. This industry has created numerous job opportunities in transport, hospitality, and construction as it encourages the upkeep of infrastructure and boosts economic growth alongside economic and cultural development (Tourism Industry: Everything you need to know about Tourism | REVFINE).

There are currently many challenges to the tourism industry. Marketing media struggles to identify and reach target audiences so adapting and keeping abreast of advancements in social media marketing platforms and following leads become critical.

Security is another key factor that affects the industry. Tourists will shy away from destinations deemed unsafe, regardless of the offered attractions and facilities of the destination in question (What Are The Top Challenges Facing the Travel Industry In 2019? | SiteVisibility 2019).

The COVID-19 pandemic resulted in global lockdowns and flight cancellations which rendered tourism locations vacant on a global scale. Restaurants and hotels incurred debilitating losses with some permanently forced to shut down. The United Nations World Tourism Organization estimated the sharp decline in tourist arrivals

© Springer Nature Switzerland AG 2023
M. Anis et al., *Mapping Innovation*,
https://doi.org/10.1007/978-3-030-93627-3_15

globally by 20–30% in 2020, leading to an estimated US$30–50 billion in losses (COVID-19: Putting People First | UNWTO).

Value Chain

This value chain can be divided into five main phases, all serving the end consumer. They are travel planning, transportation, accommodation, food and shopping, and tourist sites.

Travel planning includes researching and selecting options including transportation, accommodation, food, shopping, and visiting tourist sites, in addition to suggesting appropriate timeframe allocation. Then, tourists use selected transportation services, whether domestic or international, and pay. Afterward, the tourist checks into the chosen accommodation. Afterward, they spend time shopping for souvenirs and other needs, eating food at restaurants or preparing their food (if self-catering), and visiting touristic sites.

Case Studies

Case Study 1 Booking.com

Booking.com

- Country of implementation: Global
- Company: Booking.com

Booking.com is a travel search engine for lodging reservations owned by Booking Holdings that is headquartered in Amsterdam. It has more than 28 million listings and, as a service, is available in 43 languages. It was founded in 1996. In 2011, the company made $1.1 billion in profit, which helped it set a record of being the most profitable organization in the digital travel market.

Technology	Big data analysis/data structures, machine learning
Description	Based on user preferences such as geographical location, star rating, room price, facilities, and optional additional filters, user data is captured, recorded, and categorized using tailored software. The data is processed through a machine learning algorithm, which creates a model that can predict what users may look for in future bookings Booking.com achieved a remarkable profit from their system as its user-friendly interface makes the process of finding the best deals and hotel vacancies easier
Stage	In market
Impact on beneficiaries	Hotels • Improved hotel presence • Higher quality images • Prominent facilities and service listings Guests • User-friendly website • Access to exclusive deals • Convenient

Case Study 2

HelloGbye

| Tour Service | Transportation | Accommodation | Food & Beverage | Touristic Products |

- Countries of implementation: Global
- Company: HelloGbye

HelloGbye is an automation company aiming to enhance tourism services by enabling tourist agencies to shorten their customer response time and improve the traveler experience. It was founded by Jonathan Miller, Roy Miller, and Steve Seider in 2015 (HelloGbye 2020).

Technology	Artificial intelligence (AI)/data analytics
Description	This application enables travelers to verbally command or type their requests to receive several airfares and hotel recommendations. It also allows travelers to create profiles with several different preferences, such as ticket price, class of travel (first class, economy, etc.), single or group bookings, etc. Travelers can also create multiple profiles to distinguish preferences for their business, school trips, or family trips (Grigonis 2017)

Stage	In market	
Beneficiaries	Travelers	• Improved search results • Speedier search engine • Improved customer satisfaction • Listing/searching multiple preferences at the same time
	Travel agents	• Diversified travel options • Reduced handling time • Online process

Case Study 3 ⓥ polarsteps

Polarsteps

- Countries of implementation: Global
- Company: Polarsteps

Polarsteps is a company that aims to find new ways to track and retain traveler experience. It was founded by Job Harmsen, Koen Droste, Maximiliano Neustadt, and Niek Bokkers in 2015. In 2019, the company raised €3 million in funding (Polarsteps 2019).

Technology	Big data/Google analytics
Description	The Polarsteps application allows users to periodically check and publish their locations onto a webpage so friends and family can keep track of them Travelers can upload pictures to create multimedia albums of selected trips creating a collective shared travel experience (Kamps 2016)
Stage	In market
Beneficiaries	Guests

		• Track traveling friends and family • Record precious travel memories • Suggest and receive recommendations from fellow travelers

Case Study 4

Peek

Tour Service → Transportation → Accommodation → Food & Beverage → Touristic Products

- Countries of implementation: Global
- Company: Peek

Peek is a travel service company founded by Ruzwana Bashir in 2011 and dedicated to facilitating the booking process and allowing related businesses to stream their services and destination activities to grow their customer base.

Technology	Big data/cloud-based technology	
Description	Peek is a website that lists activities, tours, and rentals in the travelers' current destination at competitive prices (Peek 2019)	
Stage	In market	
Beneficiaries	Activity operators	• Improved revenue stream • Effective time management
	Travelers	• Quick check-ins • Customized activities at the destination • Cost-effective

Case Study 5

Hopper

- Countries of implementation: Global
- Company: Hopper

Hopper is a travel application founded by Frederic Lalonde and Joost Ouwerkerk in 2007 and currently valued at $780 million (Perez 2019).

Technology	Artificial intelligence (AI)/data analytics
Description	The Hopper application helps travelers find cost-effective flights and hotel bookings at discounts of as much as 40% of their original price. It has more than 270,000 hotels in more than 200 countries and can predict future prices with up to 95% accuracy, enabling it to recommend bookings at later dates or immediate purchase, per the client's best budgeting interests (Perez 2019)
Stage	In market
Beneficiaries	Travelers

- Cost-effective
- Improved customer satisfaction
- Best deal alerts

Case Study 6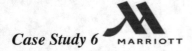

Marriott Hotels

- Country of implementation: China
- Company: Marriott Hotel

Marriott Hotel is one of the most popular multinational hospitality companies in the world with hotels in 131 countries. The Marriott Hotel chain utilizes innovative and exciting tools that provide visitors with a luxurious stay with a simple booking process.

Technology	Biometrics/recognition technology
Description	Recognition tech is utilized in Marriott Hotels in China, the Hangzhou Marriott Hotel Qianjiang, and Sanya Marriott Hotel Dadonghai Bay. Guests can check-in using facial recognition by first scanning their IDs and inputting their contact information onto a smart machine that can provide them with the room keys upon confirmation (Hong Kong 2018)
Stage	In market
Beneficiaries	Hotel guests
	Customer service

Beneficiaries		
Hotel guests	• Quick check-in time (1 minute) • Improved customer satisfaction
Customer service	• Enhanced security • Time-effective

Case Study 7

Airbnb

- Countries of implementation: Global
- Company: Airbnb

Airbnb is one of the largest marketplaces in the world to help people rent out their homes to others who are looking for accommodation. It enables homeowners to gain additional revenue streams while they are away from home. It holds millions of listings from over 220 countries and over 100,000 cities. Hosts have earned over $80 billion using Airbnb (Airbnb 2020).

Technology	Artificial intelligence (AI)/machine learning and data analytics
Description	Airbnb saves data from user preferences and uses this to suggest similar accommodations, enhancing and facilitating customer searches. It helps hosts appropriately price their properties by utilizing algorithms, including location, time of year, etc. Airbnb also determines if guests or hosts are trustworthy by performing online background checks (Marr 2020)
Stage	In market
Beneficiaries	Guests	• Price effective • Convenient
	Hosts	• Protection against fraud and property damage

Case Study 8

IBM

| Tour Service | Transportation | Accommodation | Food & Beverage | Touristic Products |

- Country of implementation: USA and Japan
- Company: IBM (International Business Machines)

IBM (International Business Machines) was founded by Charles Ranlett Flint in 1911. It uses computing, software, cloud-based, and hardware services to solve world challenging issues (Computer Business Review).

Technology	Internet of things (IoT)/Bluetooth low energy beacons	
Description	The NavCog application was designed to help people navigate the Pittsburgh International Airport. It allows travelers to locate and get to their desired destinations, including restrooms, restaurants, and gates, through voice command messages. Travelers are instructed how and where to turn at each point until they reach their target destination. The application works with Bluetooth low-energy beacons placed in the airport. This application also provides real-time updates about flight details and flight gate information to enhance the experience of the visually impaired traveler (Koines 2018)	
Stage	In market	
Beneficiaries	Visually impaired	• Improved traveler confidence • Minimizes stress of traveling • Improves trip enjoyment • Enables independent travel

Case Study 9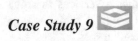

Sandblock

- Country of implementation: Global
- Company: Sandblock

Sandblock is a computer software company founded by Sarah-Diane Eck in 2017 which provides innovative ways to benefit travelers (GDI 2019).

Technology	Blockchain/crypto assets
Description	Sandblock utilizes the blockchain technology to increase customer loyalty and analyze consumer behaviors during hotel stays. Guests are awarded loyalty points which can be exchanged for *Satisfaction Tokens* and used as crypto assets to gain rewards to other services beyond those originally awarded (Liebkind 2019)
Stage	In market
Beneficiaries	Travel agents

Beneficiaries		
Travel agents		• Improved customer loyalty
Travelers		• Gain rewards

Case Study 10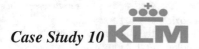

KLM Royal Dutch Airlines

Tour Service	Transportation	Accommodation	Food & Beverage	Touristic Products

- Country of implementation: Netherlands
- Company: KLM Royal Dutch Airlines

KLM Royal Dutch Airlines is one of the oldest airlines operating using its original name. In 2017, it generated €10 billion in revenue. It aims to consistently find innovative ways for travelers to have memorable experiences (KLM Royal Dutch Airlines).

Technology	Internet of things (IoT)/sensors	
Description	KLM Royal Dutch Airlines have designed smart seats called FlightBeats, equipped with electronic sensors that measure passenger heart rates and stress levels. This data is sent to flight crew tablets as a preventative measure to lessen the need for emergency land due to minor anxiety-fueled panic attacks. FlightBeats measure stress levels in real time, and the crew are subsequently able to reach the unwell patient by following a color-coded seat map on their tablets so they can employ calming methods or provide comfort (Allianz Partners Business Insights 2017)	
Stage	In market	
Beneficiaries	Passengers	• Reduced stress levels • Reduced risk of having panic attacks • Improved safety and mental health
	Flight crew	• Reduced incidences of emergency landings • Reduced risk of panic and anxiety attacks • Improved safety and mental health of passengers

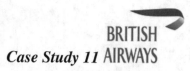

Case Study 11 BRITISH AIRWAYS

British Airways

| Tour Service | Transportation | Accommodation | Food & Beverage | Touristic Products |

- Country of implementation: UK
- Company: British Airways

British Airways is one of the largest global airlines and international transporters in the UK, flying to more than 200 destinations in 80 countries (Media Center 2019).

Technology	Internet of things (IoT)/neurosensory	
Description	The happiness blanket can measure a passenger's mood through a tech-enhanced headband fitted with neurosensory connected to passenger blankets. Blankets accordingly change color in response to the passenger's mood. By using this data, the airline can enhance the passenger experience by adjusting various factors, such as lighting, seat position, meal times, film selection, etc. (Szondy 2014)	
Stage	In market	
Beneficiaries	Passengers	• Improved relaxation levels • Reduced stress and discomfort levels
	Airline	• Improved passenger satisfaction

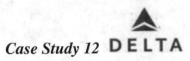

Case Study 12 **DELTA**

Delta Airlines

- Country of implementation: USA
- Company: Delta Airlines

Delta Airlines is one of the biggest airlines in the USA and transports around 200 million passengers a year to over 300 destinations in 50 countries. Delta Airlines is renowned for using innovation to enhance the passenger experience (Delta News Hub 2020).

Technology	Internet of things (IoT)/radio frequency identification (RFID)
Description	Delta Airlines utilizes RFID technology for baggage tracking. RFID scanners capture and store precise data onto RFID chips embedded in luggage tags using radio waves. Accordingly, passengers can track their bags and receive notifications throughout their journey using the Fly Delta application (Delta New Hub)

Stage	In market	
Beneficiaries	Passengers	• Updated luggage location tracking • Improved luggage handling transparency

Case Study 13

Utrip

- Country of implementation: The USA
- Company: Utrip

Utrip is a travel company founded by Gilad Berenstein and Yair Berenstein, in 2012, with a mission to ensure a heightened traveler experience of enjoyment and ease (Travel Massive).

Technology	Artificial intelligence (AI)/machine learning	
Description	Utrip uses artificial intelligence (AI) and machine learning to plan perfect trips for travelers, from hotels to activities and must-see sights. The traveler inputs their flight details and target destination and lists their preferences in 16 different categories, including history, food, budget, etc. Utrip then analyses these to determine suitable options while considering factors, such as seasonality and logistics (Travel Massive)	
Stage	In market	
Beneficiaries	Travelers	• Bespoke travel • Cost-effective • Reduced planning stress

Case Study 14 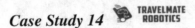 TRAVELMATE ROBOTICS

Travelmate Robotics

- Countries of implementation: USA, Europe, and Japan
- Company: Travelmate Robotics

Travelmate Robotics is an electronics and robotics company that specializes in designing and developing new ways to use robotic technology while traveling.

Technology	Artificial intelligence (AI)/robotics technology, machine learning, sensors, and Bluetooth low energy module
Description	Travelmate is the first autonomous suitcase that follows the user to a maximum speed of 6.75 mph, weighs the same as any regular suitcase, and has a built-in scale to inform the user about the weight of its contents. It has built-in sensors that allow it to avoid obstacles and is controlled using a multifeature application that includes horizontal and vertical modes (moving position of the suitcase). It allows the user to pinpoint the location of their suitcase, at all times, as the suitcase has a built-in GPS chip embedded Additionally, it has a Bluetooth-enabled lock to enhances the safety of contents. The suitcase's battery life lasts up to 4 hours and can be easily removed and charged with wireless technology. It comes in three sizes –small, medium, and large (Travelmate Robotics)
Stage	In market
Beneficiaries	Travelers

Beneficiaries	Travelers	• Convenient self-weighing luggage • Improved luggage security • Ease of charging • The convenience of hand free

Case Study 15

≡ SoftBank

Softbank Robotics

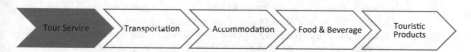

- Countries of implementation: France and Japan
- Company: SoftBank Robotics

SoftBank Robotics is a company specializing in creating and developing humanoid robots. Founded by Masayoshi Son and Ronald D. Fisher, in 2005, its robots are used in tourism, education, and health in over 70 countries (Softbank Robotics).

Technology	Artificial intelligence (AI)/robotics technology, machine learning, speech recognition, sensors, cameras	
Description	SoftBank's Pepper is a humanoid robot designed to help travelers by asking travelers a series of questions about how their day is going and their personal travel preferences. The robot then creates a traveler profile to match characteristics of the traveler's personality with destination profiles. It is available to converse in 15 different languages including English, French, Dutch, and Arabic (Arean 2018)	
Stage	In market	
Beneficiaries	Travel agents	• Extra income • Renewed business opportunities

Case Study 16 **Hilton**
HOTELS & RESORTS

Hilton

- Country of implementation: USA
- Company: Hilton Hotel and Resorts

Hilton Hotel and Resorts are a subsidiary of Hilton, an American multinational company that provides full-service hotels and resorts. The company's hotels aim to attract business travelers and tourists with hotel bookings near airports, major cities, and vacation destinations globally. It seeks to provide comfort and hospitality coupled with exceptional client experience (Business Background of Hilton Hotels | UK Essays).

Technology	Artificial intelligence (AI)/robots and speech recognition	
Description	Connie is a bot that uses AI and speech recognition to interact with guests, greeting them, and answering their questions in an informative, friendly manner. Connie is powered by IBM's Watson and WayBlazer platforms, an online travel application that uses Watson, an IBM cognitive and AI analytic software, to collect travel data from online resources (Hilton and IBM Pilot "Connie," The World's First Watson-Enabled Hotel Concierge | Hilton)	
Stage	In market	
Beneficiaries	Travelers	• Improved customer communication • Customized guest experience

Case Study 17 ATLANTIS
RESORTS & RESIDENCES

Atlantis Dubai

- Country of implementation: UAE
- Company: Atlantis

Atlantis is a luxury hotel resort in Dubai, UAE. Built on an island, with predominantly Arabian themes, the core of its character is built upon the myth of Atlantis. The resort includes a water park, a lost Atlantis-themed aquarium housing thousands of marine animals, and several high-end restaurants (Atlantis, The Palm Dubai | Circle One Studios).

Technology	Virtual reality/social media platforms	
Description	Atlantis resort has created a virtual tool to walk users through its hotel using VR glasses and a smartphone. The application provides a 360- virtual view to demonstrate the services and destination provided (Atlantis, the Palm – VR Panovideo Application | Artworks Smartphone Application)	
Stage	In market	
Beneficiaries	Travelers	• 360 pre-visit virtual experience • Cost-effective

Case Study 18

Expedia

| Tour Service | Transportation | Accommodation | Food & Beverage | Touristic Products |

- Countries of implementation: Global
- Company: Expedia

Expedia is one of the world's leading online travel agencies, providing travel products and services from a vast resource pool of airlines, car rental agencies, cooperatives, cruise lines, and much more (Expedia Group, Inc. Modern Slavery Statement | Expedia Group).

Technology	Blockchain/platform	
Description	The Winding Tree is a blockchain platform used in Expedia for baggage tracking to make it more accessible, convenient, and more secure. Winding Tree uses LIF token, a cryptocurrency token, to connect tourists with companies that provide tour guides, hotels, and airlines (6 Companies Using Blockchain to Change Travel | Investopedia)	
Stage	In market	
Beneficiaries	Travelers	• Improved foreign exchange fees • More reliable and secure payments • Easier luggage tracking

Case Study 19

United Airlines

Tour Service | Transportation | Accommodation | Food & Beverage | Touristic Products

- Country of implementation: USA
- Company: United Airlines

United Airlines dominates a significant segment of the American airline industry and is known for its comprehensive route network. Its primary goal is to serve as a bridge connecting people to unify the world. The company invests heavily in innovation to create better customer relations and enhance client experiences (Company Overview | United 2020).

Technology	Internet of things (IoT)/smart technologies and virtual assistance
Description	United Airlines uses Amazon Echo, an Amazon Alexa device, to provide clients with information on their flights. Using simple commands, users can find any trip-related information in flight. With the increase in popularity of smart speakers in the market, the airline moved forward to offer such services to clients. Guests could now "speak" to the device and know updates about the flight, check future flight schedules, know airline facilities, and check-in to their connecting flights (Introducing Hands-Free Check-In For Your Favorite Smart Device | United 2017)

Stage	In market	
Beneficiaries	Travelers	• Access to flight information • Convenient • Smooth and quick check-in

Case Study 20

Avvio

* Countries of implementation: Global
* Company: Avvio

Avvio is a booking platform that provides information on hotels and apartments. It connects different hospitality providers and encourages growth via direct booking service (About Avvio | Avvio).

Technology	Artificial intelligence/machine learning
Description	Avvio produced Allora for hotels to gain booking data and understand industry patterns. The AI technology in Allora studies customer browsing patterns then offers them customized recommendations categorized by costs, theme, location, and length of stay, helping many renowned hotels globally grow their online presence (AI in the Travel and Tourism Industry – Current Applications | Emerj)
Stage	In market
Beneficiaries	Travelers	• Enhanced booking experience • Improved online communication • Personalized online service

Glossary of Technological Terms

3D Printing

3D printing, also known as additive manufacturing, is the process of adding layers of material to build an object. Engineers and product designers upload a digital CAD tile to a 3D printer that prints it in a solid 3D form. The most frequently used material in 3D printers is thermoplastic, but the technology includes other materials such as photopolymers, epoxy resins, and metals (What is 3D Printing? | Stratasys).

Artificial Intelligence (AI)

Artificial Intelligence (AI) is the simulation of intelligent behavior in smart machines that are designed to reflect a human's mind, such as thinking, learning, communicating, problem solving, and reasoning. Multiple research papers suggest that artificial intelligence will reach a point where it will be able to think and function better than humans do. Artificial intelligence has the ability to look at different possibilities and choose the one with the best outcome. There are different concepts of artificial intelligence, such as machine learning, deep learning, data science, and many more (Nilsson 1998). Artificial intelligence is transforming the packaging industries by making the theoretical idea of smart warehouses possible. The concept of a smart warehouse allows users to collect or integrate data from different inputs in a warehouse then using this data AI can make smart decisions in the packaging processes. Using AI, machines are able to learn from patterns and adapt to new environments (Econo Corp).

© Springer Nature Switzerland AG 2023
M. Anis et al., *Mapping Innovation*,
https://doi.org/10.1007/978-3-030-93627-3

Automation

The term refers to software programs that can replace human activities that are repetitive and rule-based. The software can be implemented on a mechanical machine to form a robot that automates a certain task. Some tasks are dangerous for humans like bomb dismantling or dangerous area exploration, so robots can be supervised to do these tasks without any human intervention.

Automation and Digitalization

It is the technique of making a process operate automatically with minimal human intervention. Automation crosses all industries, from installation to maintenance and design. It consists of a broad range of technologies such as robotics, expert systems, telemetry and communication, sensors, and cybersecurity (International Society of Automation 2020).

Augmented Reality (AR)

Augmented reality (AR) is a view of the real, physical world in which users find elements enhanced by computer-generated input. This computer-generated input which is augmented on reality ranges from sound to video to graphics to GPS overlays and more, in digital content that responds in real-time to changes in the user's environment, usually movement. AR is simply a combination of real and virtual computer-generated worlds. Given a real subject, captured on video or camera, the technology augments that real-world image with extra layers of digital information (Augmented Reality 2019). The main added value of augmented reality is the way by which the components of the digital world mix into a person's perception of the real world, not as a simple display of data, but through the integration of immersive sensations which are perceived as natural parts of an environment. The applications of AR technology are endless ranging from games to education, aviation, industry, retail, marketing, and much more.

Big Data Analytics

Big data analytics is the complicated process of investigating huge data to reveal patterns, useful hidden information, and preferences to assist companies in understanding the market trends to provide better services for new revenue opportunities. Big data involves data coming from IoT sensors, online content, server logs, messages, etc. The analysis begins with organizing these unstructured amounts of data with software; then, data scientists write machine learning algorithms, which are blocks of codes based on statistical and mathematical models that can predict future behavior based on historical data sets to help big supply chains to improve decision-making.

Biometrics

Biometrics are biological measurements that can be used to identify someone. It includes fingerprint mapping, retina scans, facial recognition, voice recognition, and others. Some applications of biometrics also include monitoring someone's heartbeat in order to verify their identity. Biometric scanners are being widely used for several purposes, including security and identification (Kaspersky). Biometrics today has a huge impact on businesses and industries. Businesses use biometrics in order to use eye-tracking and facial expressions technology so that they can collect information and data about the consumer's reactions on the packaging and labels. They are able to tell how long the consumer looked at the product for and if it was exciting enough or not (Dollard Packaging).

Blockchain

Blockchain technology is a chain of digital information about transactions stored globally across thousands of network computer systems. Blockchain technology is a type of distributed ledger technology (DLT); therefore, each block that has digital information is stored with a unique cryptographic code created by algorithms, called a hash, which identifies it from the other blocks (Mearian 2019). One of the biggest contributors in the packaging industries is the blockchain technology. It allows users to track and trace their packages, facilitates customer awareness, and protects brands from fake products. Using the blockchain technology, customers can easily see the journey of the package from the production till reaching the end user, and this increases engagement between the package and the customer (Shaw 2019).

Green Technologies

It is the quality of air, water, waste, soil, and others. Green technology is by definition environmentally friendly. The different fields or topics that fall under green technology are health concerns, recycling, and renewable resources; it tries to reduce waste, conserve water, reduce carbon emission, and conserve energy (Bhardwaj and Neelam 2015). Using green technologies in industries, users will be able to design and product packaging products that are sustainable and environmentally friendly. Packaging products that are environmentally friendly should be designed with a very light and recyclable material that does not produce high carbon emissions (Wright).

Internet of Thing (IoT)

Internet of things refers to the numerous numbers of physical computing devices that are connected to the internet and collecting and sharing data. All these devices have the ability to connect to one another and add special sensors in order to send or share real-time data without involving a human being. In other words, internet of things is made up of devices that are connected together with automated systems in order to gather information or data, analyze it, and create an action. Internet of things evolved from the internet to wireless connection and from micro-electromechanical systems to embedded systems. The different fields that support IoT are control systems, wireless sensor networks, automation and many more (Foote 2016). Internet of things is changing the packaging industry greatly by creating innovative ways for businesses to easily collect real-time data and information about the machine and measure its performance. Industries will be able to monitor all machines and immediately track and trace any fault or malfunction within seconds (Industrial IoT 2017).

Machine Learning (ML)

Machine learning is a branch of artificial intelligence where the system can self-learn, recognize patterns, and make decisions without being explicitly programmed. There are two classes of methods used in ML, which are supervised and unsupervised algorithms. The supervised algorithms depend on having data with labels on them, so the system is responsible for finding the mapping function so that it can predict the result when it has new data. In unsupervised learning, we do not have labels on the data, so the system is responsible for finding the hidden structure within the data such as the clustering problems (Supervised and Unsupervised Machine Learning Algorithms | Machine Learning Mastery 2019).

Nanotechnology

Nanotechnology is science and technology conducted at the nanoscale, typically in the range of one to a hundred nanometers. Nanotechnology in medicine involves applying nanoparticles in different medical pursuits such as medicine and making repairs at the cellular level (What is Nanotechnology | National Nanotechnology Initiative).

Virtual Reality (VR)/Augmented Reality (AR)

VR is the creation of a simulated environment using computer technology. The head-mounted display HMD is one of the most known display technologies that can put human beings in simulated interfaces where they can interact with 3D objects, unlike usual 2D screens. On the other hand, AR simulates artificial objects in the real surrounding environment by using sensors and algorithms to superimpose computer-generated images on the user's real view (What is Virtual Reality | Marxent 3D Commerce 2019).

Works Cited

(2014) Ekso Bionics™ and SoldierSocks Expand Partnership With Three-Year, 80-Unit Pledge. In: Ekso Bionics. https://ir.eksobionics.com/press-releases/detail/324/ekso-bionicstm-and-soldiersocks-expand-partnership-with. Accessed 1 July 2020.

(2016) Levi's Real-Time Tracking of Jeans: RFID in Retail. In: RTInsights. https://www.rtinsights.com/rfid-in-retail-customer-experience-levis/. Accessed 22 July 2020.

(2017) 5 Benefits: Competitive Advantages of big data in Business In newgenapps. https://www.newgenapps.com/blog/importance-benefits-competitive-advantage-big-data/. Accessed 24 June 2020.

(2017) 5 Medical Robots Making a Difference in Healthcare In Case Western Reserve University. https://online-engineering.case.edu/blog/medical-robots-making-a-difference. Accessed 16 July 2020.

(2017) Augmedix: Humanizing Healthcare Through Google Glass. In: Digital Initiative. https://digital.hbs.edu/platform-digit/submission/augmedix-humanizing-healthcare-through-google-glass/. Accessed 16 July 2020.

(2017) Carnegie Learning Partners with OpenStax to Offer the Most Powerful and Affordable College Math Solution on the Market. In: Business Wire. https://www.businesswire.com/news/home/20171116005119/en/Carnegie-Learning-Partners-OpenStax-Offer-Powerful-Affordable. Accessed 9 July 2020

(2017) IKEA Launches IKEA Place, a New App That Allows People to Virtually Place Furniture in Their Homes. In: IKEA. https://newsroom.inter.ikea.com/news/ikea-launches-ikea-place%2D%2Da-new-app-that-allows-people-to-virtually-place-furniture-in-their-home/s/f5f003d7-fcba-4155-ba17-5a89b4a2bd11. Accessed 22 July 2020.

(2017) Mixed Reality in Healthcare - The HoloLens Review. In: The Medical Futurist. https://medicalfuturist.com/mixed-reality-healthcare-hololens-review/. Accessed 1 July 2020.

(2018) 10 Challenges and Opportunities Media and Entertainment Industry in 2018. In: LinkedIn. https://www.linkedin.com/pulse/10-challenges-opportunities-media-entertainment-2018-janet-balis/. Accessed 24 June 2020.

(2018) Chatbots in Retail: Nine Companies Using AI to Improve Customer Experience. In: Real Insight Network. https://www.retail-insight-network.com/features/chatbots-in-retail-ai-experience/. Accessed 22 July 2020.

(2018) Worldwide Adoption of Ekso Bionics EksoGT Exoskeleton Allows Stroke and Spinal Cord Injury Patients to Take 100 Million Steps to Date. In: Globe Newswire. https://www.globenewswire.com/news-release/2018/12/04/1661768/0/en/Worldwide-Adoption-of-Ekso-Bionics-EksoGT-Exoskeleton-Allows-Stroke-and-Spinal-Cord-Injury-Patients-to-Take-100-Million-Steps-to-Date.html. Accessed 1 July 2020.

© Springer Nature Switzerland AG 2023
M. Anis et al., *Mapping Innovation*,
https://doi.org/10.1007/978-3-030-93627-3

(2019) Blockcertś in Bestr: FAQ. In: Bestr Blog. https://blog.bestr.it/en/2019/06/13/blockcerts-bestr-faq. Accessed 9 July 2020.

(2019) Media and Advertising are Changing. Here are your Biggest Challenges. In: Inc. https://www.inc.com/tracy-leigh-hazzard/media-advertising-is-changing-here-are-your-biggest-challenges.html. Accessed 24 June 2020.

(2019) Nanotechnology in Medicine: Who Are the Leading Public Companies. In: Verdict Medical Device. https://www.medicaldevice-network.com/comment/nanotechnology-in-medicine-who-are-the-leading-public-companies/. Accessed 16 July 2020.

(2020a) What Is Automation. In: Techopedia. https://www.techopedia.com/definition/32099/automation. Accessed 31 July 2020.

(2020b) What Is Automation. In: Techopedia. https://www.techopedia.com/definition/32099/automation. Accessed 31 July 2020.

3D printing, "What is 3D Printing", Retrieved from: https://3dprinting.com/what-is-3d-printing/

A Brief History of Microsoft - The World's Biggest Software Company. In: DSP. https://content.dsp.co.uk/a-brief-history-of-microsoft-the-worlds-biggest-software-company. Accessed 1 July 2020.

A Brief History of Microsoft - The Worlds Biggest Software Company. In: DSP.

A Google Spinoff is Developing First Pokémon Game for smartphones. In: Fortune. https://fortune.com/2015/12/21/pokemon-ingress-niantic-labs/. Accessed 24 June 2020.

A.T.E. Group, "Online Monitoring & Analytics for Weaving", Retrieved from https://www.ategroup.com/wp-content/uploads/brochures/EcoAxis_weaving-brochure-Apr-18.pdf

About Amgen. In: Amgen. https://www.amgen.com/about/. Accessed 16 July 2020.

About ASCO. In: Cancer.Net. https://www.cancer.net/about-us/about-asco. Accessed 16 July 2020.

About Ciox. In: Ciox. https://www.cioxhealth.com/about-us. Accessed 16 July 2020.

About Da Vinci Systems. In: Intuitive. https://www.davincisurgery.com/da-vinci-systems/about-da-vinci-systems#. Accessed 1 July 2020.

About DAQRI. In: DAQRI. https://daqri.com/about/. Accessed 9 July 2020

About EnvisionTEC. In: EnvisionTEC. https://envisiontec.com/company/?gclid=Cj0KCQjw9b_4BRCMARIsADMUIyq1l4ZU2fp3Fuy_SQHq3BUNsELFfWyFnf_k2OHdIBFYFpgAiarX8T8aAurGEALw_wcB. Accessed 16 July 2020.

About Green Hippo. In: Green Hippo. https://www.green-hippo.com/about/. Accessed 24 June 2020.

About Intuitive. In: Intuitive. https://www.intuitive.com/en-us/about-us/company. Accessed 1 July 2020.

About Netflix. In: Netflix Media Center. https://media.netflix.com/en/about-netflix. Accessed 24 June 2020.

About Nike. In: Nike. https://about.nike.com. Accessed 22 July 2020.

About Smart Technologies. In: Smart Technologies. https://www.smarttech.com/en/about. Accessed 25 April 2020

About the Author: "The State of Biometrics Solutions: Use Cases and Advances." Aware, 20 Jan. 2020, www.aware.com/blog-state-of-biometrics-solutions/.

About Us. In: Bose. https://www.bose.com/en_us/about_bose.html. Accessed 24 June 2020.

About Us. In: CaliBurger. https://caliburger.com/about-us. Accessed 22 July 2020.

About Us. In: Ekso Bionics. https://eksobionics.com/company/about-us/. Accessed 1 July 2020.

About Us. In: Pfizer. https://www.pfizer.co.uk/about-us. Accessed 16 July 2020.

About. In: Cloudera https://www.cloudera.com/about.html. Accessed 27 April 2020

About. In: Moodle. https://moodle.com/about/. Accessed 29 April 2020

About. In: Zoom. https://www.zoom.us/about Accessed 28 April 2020. Accessed 29 April 2020

Accessed 1 July 2020.

Accessed 31 July 2020.

Accuray Incorporated. In: Reuters. https://www.reuters.com/companies/ARAY.O. Accessed 16 July 2020.

Actility (2020) Logistics and Supply Chain Additive Manufacturing, "Betatype Begins Serial Production for 3D Printed Automotive Headlight Heatsink", Retrieved from: https://additivemanufacturingtoday.com/betatype-begins-serial-production-for-3d-printed-automotive-headlight-heatsink

Advantages and Disadvantages of Pro Tools. In: HowStuffWorks. https://entertainment.howstuffworks.com/pro-tools-software-hardware2.htm. Accessed 24 June 2020.

Ahmed, Faran, et al. "ICT and Renewable Energy: a Way Forward to the Next Generation Telecom Base Stations." Telecommunication Systems, Springer US, 1 Jan. 1970, link.springer.com/article/10.1007/s11235-016-0156-4.

Alex. (2019, May). BMW Delivers IoT Services to 1M Car Owners by Using IBM Cloud Foundry. Retrieved from Altoros: https://www.altoros.com/blog/bmw-delivers-iot-services-to-1m-car-owners-by-using-ibm-bluemix/

Alexander, A. (2016), "Green Technology – Disruptive Trends Transforming the Auto Industry", Retrieved from: https://usgreentechnology.com/green-technology-disruptive-auto-trends/

Amazon. In: Wikipedia. https://en.wikipedia.org/wiki/Amazon_(company). Accessed 22 July 2020.

Amefrid (2020a) Expect More An Overview of Retailing. http://www.pondiuni.edu.in/storage/dde/downloads/markiii_rm.pdf. Accessed 22 July 2020.

Amefrid (2020b) Overview An Overview of Retailing. http://www.pondiuni.edu.in/storage/dde/downloads/markiii_rm.pdf. Accessed 22 July 2020.

Andrew Hawkins. (2018a, Jan). Toyota's 'e-Palette' is a weird, self-driving modular store on wheels. Retrieved from TheVerge: https://www.theverge.com/2018/1/8/16863092/toyota-e-palette-self-driving-car-ev-ces-2018

Anna B., 2020. Peerform review: Peer-to-peer loans with fees that can add up. Retrieved from https://www.creditkarma.com/personal-loans/i/peerform-review-personal-loans

AR and VR - A Game Changer in the Automotive Industry. Retrieved from AxisCades: https://www.axiscades.com/ar-and-vr.html

Archer, "How is 3D Printing Used in the Automotive Industry", Retrieved from: https://archersoft.com/blog/how-3d-printing-used-automotive-industry

Arjun Singh. In: . https://www..com/person/arjun-singh-2#section-overview. Accessed 9 July 2020

Artthra 2020. https://arttha.com/digital-banking/

Arvind (2020a) About Arvind https://www.newindianexpress.com/topic/Arvind_Kejriwal

Arvind (2020b) Energy Arvind https://www.newindianexpress.com/topic/Arvind_Kejriwal

ASOS. In: Wikipedia. https://en.wikipedia.org/wiki/ASOS_(retailer). Accessed 22 July 2020.

Augmented Reality. (2019). Retrieved from Interaction Design Foundation: https://www.interaction-design.org/literature/topics/augmented-reality

Automotive Alliance, "Automotive Business Opportunities", Retrieve from: https://northernautoalliance.com/services/automotive-business-opportunities/

Automotive Mastermind. (2017, Jul). Volkswagen Dealerships Can Now Partner with Predictive Analytics Technology Company to Boost Sales and Customer Retention Rates. Retrived from Cision PR Newswire: https://www.prnewswire.com/news-releases/volkswagen-dealerships-can-now-partner-with-predictive-analytics-technology-company-to-boost-sales-and-customer-retention-rates-300491772.html

Automotive UX, "Car2Go with a Beemer: BMW Reach Now", Retrieved from: http://www.automotive-ux.com/news/bmw-reachnow/

Avid Acquires the Assets of Rocket Networks. In: BusinessWire. https://www.businesswire.com/news/home/20030404005511/en/Avid-Acquires-Assets-Rocket-Networks-New-Technology. Accessed 24 June 2020.

Ayan D., 2019. Robo-advisors and Artificial Intelligence – Comparing 5 Current Apps. Retrieved from https://emerj.com/ai-application-comparisons/robo-advisors-artificial-intelligence-comparing-5-current-apps/

Baird N. (2019) The Four Definitive Use Cases For AR and VR in Retail. In: Forbes. https://
www.forbes.com/sites/nikkibaird/2019/03/25/the-four-definitive-use-cases-for-ar-and-vr-in-
retail/#127ed4c969c2. Accessed 22 July 2020.

Banerjee, Ari. "Big Data & Advanced Analytics in Telecom: A Multi-Billion Revenue Opportunity."
Huawei, 2016a, www.huawei.com/ilink/en/download/HW_323807.

Bardi J (2019a) What is Virtual Reality. In: Marxent 3D Commerce. https://www.marxentlabs.
com/what-is-virtual-reality/. Accessed 16 July 2020.

Bardi J., 2019b What is Virtual Reality? [Definition and Examples]. Retrieved from https://www.
marxentlabs.com/what-is-virtual-reality/

Bernard Marr. (2017, Jul). How BMW And Parkmobile Are Using Big Data And IoT To Change The Way
We Park Our Cars. Retrieved from Forbes: https://www.forbes.com/sites/bernardmarr/2017/07/21/
how-bwm-and-parkmobile-are-changing-the-way-we-park-our-cars/#4520ce682b08

Bernards M., 2020. The Top PropTech Trends: 6 Technologies Disrupting The Property And
Real Estate Industry. Retrieved from https://www.forbes.com/sites/bernardmarr/2020/02/03/
the-top-proptech-trends-6-technologies-disrupting-the-property-and-real-estate-
industry/#55e2acf3dc16

Betatype, "A Powerful and Flexible Technology Stack for Metal AM", Retrieved from: https://
www.betaty.pe

Bharadwaj L., 2020. Technology in Banking: 10 Innovations That Will Impact Future of Banking.
Retrieved from https://www.wowso.me/blog/technology-in-banking

Bharadwaj, R. (2019) "Artificial Intelligence in the Textile Industry – Current and
Future Applications", Retrieved from https://emerj.com/ai-sector-overviews/
artificial-intelligence-in-the-textile-industry-current-and-future-applications/

Bhardwaj M. and Neelam M., "The Advantages and Disadvantages of Green Technology", Journal
of Basic and Applied Engineering Research, p-ISSN: 2350-0077; e-ISSN: 2350-0255; Volume
2, Issue 22; pp. 1957–1960, 2015a

Bhardwaj M. and Neelam M., The Advantages and Disadvantages of Green Technology, Journal of
Basic and Applied Engineering Research, p-ISSN: 2350-0077; e-ISSN: 2350-0255; Volume 2,
Issue 22; pp. 1957–1960, 2015b https://www.krishisanskriti.org/vol_image/21Jan201607013
9zzzzzzzzzzzzzzzzzzzzzzzzzzzzzzzzzzzzzz04%20%20Monu%20Bhardwaj%20_Applied%20
1957-.pdf

Big Data Analytics. In: IBM. https://www.ibm.com/eg-en/analytics/hadoop/big-data-analytics.

Biz Intellia, "Application of IoT in Automotive Industry | Future of Automobiles", Retrieved from:
https://www.biz4intellia.com/blog/iot-applications-in-automotive-industry/

Black Mirror: Bandersnatch' Could Signal the future for A. I Directors. In: Inverse. https://www.
inverse.com/article/52470-black-mirror-bandersnatch-ai-directors. Accessed 24 June 2020

Blackburn J (2019) A&E Releases 100% Recycled Thread Line Eco100

Blippar. https://www.blippar.com/blog/2020/04/07/bringing-the-augmented-reality-education-to-
your-home-for-free. Accessed 29 April 2020

BMW, " BMW Connected", Retrieved from: https://www.bmw.com.sg/content/dam/bmw/
common/topics/offers-and-services/bmw-connecteddrive-forusers/user-instructions/BMW_
ConDrive_HowTo_Guide_CDServices_BMWConnected_EN.pdf.asset.1477659819972.pdf

Bowles R. (2020) Warehouse Robotics: Everything You Need to Know in 2019. In: Logiwa. https://
www.logiwa.com/blog/warehouse-robotics. Accessed 22 July 2020.

Brainly. In: Wikipedia. https://en.wikipedia.org/wiki/Brainly. Accessed 9 July 2020

Bringing the Augmented Reality Education to Your Home For Free (2020). In:

Brownlee J (2016) Supervised and Unsupervised Machine Learning Algorithms. In: Machine
Learning Mastery. https://machinelearningmastery.com/supervised-and-unsupervised-
machine-learning-algorithms/. Accessed 31 July 2020.

Burch, Aaron. "The Top 10 Companies Working on Education in Virtual Reality and
Augmented Reality." Touchstone Research, 29 Sept. 2016a, touchstoneresearch.com/
the-top-10-companies-working-on-education-in-virtual-reality-and-augmented-reality/.

Burns, M. (2020), "Top 5 Challenges Automotive Industry Will Face in the 2020s", Retrieved from: https://www.digitalistmag.com/digital-economy/2020/01/14/challenges-automotive-industry-is-about-to-face-in-2020s-06202269/

Campuses Have Used IoT Projects to Promote Convenience, a., Learning, H., Upgrades, K., Schools, N., Secure, T., & Virtually, H. et al. (2021). Assistive Technology for Students with Disabilities - The Tech Edvocate. Retrieved 19 January 2021, from https://www.thetechedvocate.org/assistive-technology-students-disabilities/

Car Connectivity Services by Borgward & Orange. Ericsson.com, 9 Mar. 2020a, www.ericsson.com/en/cases/2019/orange-and-borgward.

Carol M., 2017a. Big Data Opportunities for Telecommunications. Retrieved from https://mapr.com/blog/big-data-opportunities-telecommunications/

Carrie Hampel. (2020, March). BMW uses blockchain to increase resource transparency. Retrieved from Electrive: https://www.electrive.com/2020/03/31/bmw-uses-blockchain-for-purchase-transparency/

Case Study: How Walmart Brought Unprecedented Transparency to the Food Supply Chain With Hyperledger Fabric. In: Hyperledger. https://www.hyperledger.org/learn/publications/walmart-case-study. Accessed 22 July 2020.

Chao G, Cheung J, Haller K, Lee J. The Coming AI Revolution in Retail and Consumer Products. In: IBM. https://www.ibm.com/downloads/cas/NDE0G4LA. Accessed 22 July 2020.

Chatfield, C. (2014), "How Food Manufacturers can Benefit from Going Green", Retrieved from https://www.nist.gov/blogs/manufacturing-innovation-blog/how-food-manufacturers-can-benefit-going-green

Chen D. (2018) Augmented Reality + Retail: A Glimpse in the Future of Shipping. In: Bazaar Voice. https://www.bazaarvoice.com/blog/ar-augmented-reality-shopping-retail/. Accessed 22 July 2020.

Christiansen B(2019) The Biggest Beneficiaries of Predictive Maintenance

CIB, 2019. https://www.cibeg.com/Arabic/Personal/Cards/Pages/CIB-Meeza-Cards.aspx

Cloud Collaboration. In: SearchCloudComputing. https://searchcloudcomputing.techtarget.com/definition/cloud-collaboration. Accessed 24 June 2020.

Cloudera Reports First Quarter Fiscal 2021 Financial Results. (2021). Retrieved 19 January 2021, from https://investors.cloudera.com/news-and-events/news/press-release-details/2020/Cloudera-Reports-First-Quarter-Fiscal-2021-Financial-Results/default.aspx

COGNEX (2020a) About COGNEX

COGNEX (2020b) Textile Inspection Industry Overview

Company. In: Integra Sources. https://www.integrasources.com/company/. Accessed 26 April 2020

Comstock J (2015) Virtual Nurse Startup Sense, ly Raises $2.2M, Sets Sights On Global Rollout. In: Mobile Health News. https://www.mobihealthnews.com/44577/sense-ly-raises-2-2m-sets-sights-on-global-rollout. Accessed 1 July 2020.

Connor C., 2020a. How are Telecoms Using the Internet of Things (IoT)? https://www.sdxcentral.com/5g/iot/definitions/telecom-using-iot/

Content Technologies, Inc. In . https://www..com/organization/content-technologies-inc-cti#section-overview. Accessed 9 July 2020

Cooper Tire completes guayule grant (2017, Aug). Retrieved from Rubber and Plastics News: https://www.rubbernews.com/article/20170810/NEWS/170819996/cooper-tire-completes-guayule-grant

Counter Point Research, "Weekly Updates COVID-19 Impact Global Automotive Industry", Retrieved from: https://www.counterpointresearch.com/weekly-updates-covid-19-impact-global-automotive-industry/

Crouch H (2018) Babylon Partners Up With Bupa For 'One of a Kind' Health Service. In: Digital Health. https://www.digitalhealth.net/2018/06/exclusive-babylon-bupa-partnership/. Accessed 1 July 2020.

Crowd Analyzer, 2020. Retrieved from https://www.crowdanalyzer.com

,   "Moley   Robotics",   Retrieved   from   https://www..com/organization/
        moley-robotics#section-overview

,   "Moley   Robotics",   Retrieved   from   https://www..com/organization/
        moley-robotics#section-overview

,   "Noble   Biomaterials",   Retrieved   from   https://www..com/organization/
        noble-biomaterials-inc#section-overview

,   "Olam   International",   Retrieved   from   https://www..com/organization/
        olam-international#section-overview

,   "Olam   International",   Retrieved   from   https://www..com/organization/
        olam-international#section-overview

,   "Olympus   Automation",   Retrieved   from   https://www..com/organization/
        olympus-automation#section-overview

,   "Olympus   Automation",   Retrieved   from   https://www..com/organization/
        olympus-automation#section-overview

,   "Outlast   Technologies   Inc",   Retrieved   from   https://www..com/organization/
        outlast-technologies-inc

, "Persado", Retrieved from https://www..com/organization/persado

, "Sewbo", Retrieved from https://www..com/organization/sewbo#section-overview

,   "SoftWear   Automation",   Retrieved   from   https://www..com/organization/
        softwear-automation#section-overview

,   "Toray   Industries",   Retrieved   from   https://www..com/organization/
        toray-industries#section-overview

,   "Alliance   Trading",   Retrieved   from   https://www..com/organization/
        alliance-trading#section-overview

,   "Alliance   Trading",   Retrieved   from   https://www..com/organization/
        alliance-trading#section-overview

, "BMW", Retrieved from: https://www..com/organization/bmw#section-overview

, "Google", Retrieved from: https://www..com/organization/google

, "Marelli", Retrieved from: https://www..com/organization/marelli#section-overview

, "Mercedes-Benz", Retrieved from: https://www..com/organization/mercedes-benz

, "Microsoft", Retrieved from: https://www..com/organization/microsoft#section-overview

,   "Mitsubishi   Electric",   Retrieved   from: https://www..com/organization/
        mitsubishi-electric#section-overview

, "National Highway Traffic Safety Administration", Retrieved from: https://www..com/
        organization/national-highway-traffic-safety-administration#section-overview

, "Parkmobile", Retrieved from: https://www..com/organization/parkmobile#section-overview

, "Reach Now", Retrieved from: https://www..com/organization/reachnow#section-overview

, "Siemens", Retrieved from: https://www..com/organization/siemens#section-overview

, "Toyota", Retrieved from: https://www..com/organization/toyota#section-overview

, "VEO Robotics", Retrieved from: https://www..com/organization/veo-robotics#section-overview

Daniel F., 2019. Machine Learning in Real Estate – Trends and Applications
        Presented.   Retrieved   from   https://emerj.com/ai-sector-overviews/
        machine-learning-in-real-estate-trends-and-applications/

Datatex (2020a) About Datatex retrieved from https://datatex.com/company-profile/ Accessed 19
        May 2020

Datatex (2020b) Software for Purchase Order Management Retrieved from https://datatex.com/
        the-future-of-textile-industry/ Accessed 30 June 2020.

Develco Products, 2020. Intelligent Building Automation. Retrieved from https://www.develco-
        products.com/business-areas/smart-home/

Digiteum, 2019. IoT Solutions for agricultural Irrigation System. Retrieved from https://www.
        digiteum.com/iot-solutions-agricultural-irrigation-system

Dolby Atmos: What is it? How Can You Get it? In: WhatHiFi. https://www.whathifi.com/advice/
        dolby-atmos-what-it-how-can-you-get-it. Accessed 24 June 2020.

Dorri M, Yarmohammadian M, Nadi M (2012). A Review on Value Chain in Higher Education. In: Sciencedirect. https://www.sciencedirect.com/science/article /pii/S1877042812018939. Accessed 23 April 2020

Eduqas, "Biometrics", Retrieved from http://resource.download.wjec.co.uk.s3.amazonaws.com/ vtc/2016-17/16-17_1-4/website/category/2/smart-fibres/biometrics.html

Eichenberg, P. (2020), "5 Ways Blockchain is Changing the Automotive Industry", Retrieved from: https://www.qad.com/blog/2020/03/5-ways-blockchain-is-changing-the-automotive-industry

Embodied Labs. In: LinkedIn. https://www.linkedin.com/company/embodiedlabs/about/. Accessed 16 July 2020.

EnvisionTEC Dental 3D Printers: Accurate, Fast, Reliable, And Flexible. In: EnvisionTEC. https:// envisiontec.com/3d-printing-industries/medical/dental/. Accessed 16 July 2020.

Facebook to Acquire Oculus. In: Facebook. https://about.fb.com/news/2014/03/facebook-to-acquire-oculus/. Accessed 23 May 2020.

Fagella D. (2020) Artificial Intelligence in Retail -10 Present and Future Use Case. In: Emerj. https://emerj.com/ai-sector-overviews/artificial-intelligence-retail/. Accessed 22 July 2020.

Faggella D. (2019) Examples of Artificial Intelligence in Education. In: Emerj. https://emerj.com/ ai-sector-overviews/examples-of-artificial-intelligence-in-education/. Accessed 9 July 2020

FAO (2020) COVID-19 Global Economic Recession

Fiber2Fashion (2014) Textile Sector Gets Future Ready

Finance Monthly, 2020. The Impact Of COVID-19 On The Real Estate Industry. Retrieved from https:// www.finance-monthly.com/2020/05/the-impact-of-covid-19-on-the-real-estate-industry/

Find Biometrics, "Biometric Tech is Coming to the Automotive Industry in a Big Way", Retrieved from: https://findbiometrics.com/biometric-tech-automotive-industry-503276/

Finextra, 2020. Finastra welcomes more apps to FusionStore marketplace. https://www.finextra. com/pressarticle/81613/finastra-welcomes-more-apps-to-fusionstore-marketplace

Foote K., A Brief History of the Internet of Things, Dataversity, 2016 https://www.dataversity.net/ brief-history-internet-things/#

For Better Business Just Add Pepper. In: SoftBank Robots. https://softbankrobotics.com/us/pepper. Accessed 22 July 2020.

Foursource (2020) About Foursource

GarmentIo (2020)

GE Digital. In: Wikipedia. https://en.wikipedia.org/wiki/GE_Digital. Accessed 22 July 2020.

Georgetown University provides the entire campus with cost-effective lynda.com instruction (2020). In: LinkedIn Learning. https://learning.linkedin.com/content/dam/me/learning/case-studies/ldc-casestudy-georgetown-university.pdf. Accessed 28 April 2020

GEP Smart (2020a) About GEP Smart

GEP Smart (2020b) Inventory Management Software

Glassdoor, 2020. Circle Internet Financial. Retrieved from https://www.glassdoor.com/Overview/ Working-at-Circle-Internet-Financial-EI_IE1291298.11,36.htm

Gleb B., 2020. Five Innovative Ways You Can Use Virtual Reality in the Real Estate Business. Retrieved from https://rubygarage.org/blog/virtual-reality-in-real-estate

Global News, "Automation Industry Amid Global COVID-19 Coronavirus Crisis Meticulous Research Viewpoint", Retrieved from: https://www.globenewswire.com/ news-release/2020/05/12/2031882/0/en/Automation-Industry-Amid-Global-COVID-19-Coronavirus-Crisis-Meticulous-Research-Viewpoint.html

Google Cloud (2020) Hermin Textile Google. In: Wikipedia https://en.wikipedia.org/wiki/Google. Accessed 30 April 2020

Gottsegen G (2019) Can Virtual Reality Change The Way We Think About Health? In: Builtin. https://builtin.com/healthcare-technology/ar-virtual-reality-healthcare. Accessed 16 July 2020.

Green Hippo Solutions. In: Green Hippo. https://www.green-hippo.com/about/. Accessed 23 May 2020.

Gross E. (2018) 4 Ways Artificial Intelligence is Revolutionizing Education. In: Dell Technologies. https://www.delltechnologies.com/en-us/perspectives/4-ways-artificial-intelligence-is-revolutionizing-education/. Accessed 9 July 2020

Grote Company, "Pizza Topping Line", Retrieved from https://www.grotecompany.com/en-us/Pizza-Topping-Line

Guilford Manufacturing, "There Has Been A lot of Noise on Cryptocurrencies and Bitcoin of Late", Retrieved from https://www.gulfoodmanufacturing.com/news%2D%2Dtrends/how-blockchain-technology-could-transform-the-food-industry#/

H&M. In: Wikipedia. https://en.wikipedia.org/wiki/H%26M. Accessed 22 July 2020.

Hart M (2019) RevDesinfectie Robots Deploys Xenex LightStrike Robots for Contamination Control in Pharmaceutical Cleanrooms. In: Business Wire. https://www.businesswire.com/news/home/20190409005763/en/RevDesinfectie-Robots-Deploys-Xenex-LightStrike-Robots-Contamination. Accessed 16 July 2020.

Hartmans A. (2017) I Learned How To Apply Makeup Using A Futuristic New Feature on Sephora's App - Here's What Happened. In: Business Insider. https://www.businessinsider.com/sephora-visual-artist-app-feature-teaches-how-to-apply-makeup-using-ai-photos-2017-3. Accessed 22 July 2020.

Hawkins, A.J. (2018b), 'Toyota's 'e-Palette' is a Weird, Self-Driving Modular Store on Wheels", Retrieved from: https://www.theverge.com/2018/1/8/16863092/toyota-e-palette-self-driving-car-ev-ces-2018

Hermin (2020) About Hermin https://cloud.google.com/customers/hermin-textile Accessed 17 May 2020

Heuritech. In: LinkedIn. https://www.linkedin.com/company/heuritech/. Accessed 31 July 2020.

Hexoskin, "Hexoskin Smart Garments Specifications", Retrieved from https://www.hexoskin.com

Hippotizer. In: Lexair. https://www.lexair.com.au/pages/hippotizer. Accessed 24 June 2020.

How is Technology Transforming the Media and Entertaining Industry. In: Jellyfish Technologies. https://www.jellyfishtechnologies.com/how-is-technology-transforming-the-media-and-entertainment-industry.html. Accessed 24 June 2020.

How It Works. In: Smart Attendance Solution. http://www.smartattendancesolution.com/#HowItWorks. Accessed 9 July 2020

https://content.dsp.co.uk/a-brief-history-of-microsoft-the-worlds-biggest-software-company.

https://emerj.com/ai-sector-overviews/artificial-intelligence-in-the-textile-industry-current-and-future-applications/ Accessed 17 May 2020

https://garment.io/ Accessed 19 May 2020

https://industrytoday.com/the-biggest-beneficiaries-of-predictive-maintenance/

https://new.abb.com/news/detail/61772/how-artificial-intelligence-is-revolutionizing-the-food-and-beverage-industry

https://new.abb.com/news/detail/61772/how-artificial-intelligence-is-revolutionizing-the-food-and-beverage-industry

https://reliefweb.int/report/world/covid-19-global-economic-recession-avoiding-hunger-must-be-centre-economic-stimulus-24 Accessed 15 May 2020

https://retail.economictimes.indiatimes.com/news/apparel-fashion/apparel/arvind-plans-to-cut-carbon-emissions-by-30-pc/67977474 Accessed 18 May 2020

https://textilelearner.blogspot.com/2014/08/current-challenges-in-global-textile.html Accessed 15 May 2020

https://usa.kaspersky.com/resource-center/definitions/biometrics

https://www.actility.com/ Accessed 19 May 2020

https://www.actility.com/logistics-supply-chain/

https://www.arvind.com/about-us Accessed 17 May 2020

https://www.arvind.com/energy Accessed 17 May 2020

https://www.cognex.com/company Accessed 17 May 2020

https://www.cognex.com/downloads/literaturemain?event=90f4e852-2645-47bb-b267-1baa16ab25f8 Accessed 17 May 2020 https://www.fibre2fashion.com/industry-article/7397/textile-sector-gets-future-ready Accessed 15 May 2020

https://www.foursource.com/about-us/ Accessed 19 May 2020

https://www.frontiersin.org/articles/10.3389/fdata.2018.00006/full#B6 Accessed 16 May 2020

https://www.hermin.com/about.php Accessed 16 May 2020

https://www.investopedia.com/terms/g/green_tech.asp Access May 16, 2020

https://www.investopedia.com/terms/i/internet-things.asp Accessed 15 May 2020

https://www.jeanologia.com/aboutjeanologia/ Accessed 18 May 2020

https://www.jeanologia.com/portfolio/eim-environmental-impact-software/   Accessed   19 May 2020

https://www.loriot.io/blog/IoT-trends-2020.html Accessed 15 May 2020

https://www.persado.com/about/ Accessed 17 May 2020

https://www.smartbygep.com/company/about-us Accessed 19 May 2020

https://www.smartbygep.com/procurement-software/direct-procurement-technology/ inventory-management-software

https://www.techopedia.com/definition/3411/platform-computing   Accessed 16 May 2020

Hunt, M. (2019), "Automotive Mastermind New Sales Platform to Increase Dealership Sales", Retrieved from: https://www.onlinemarketplaces.com/articles/23347-automotivemastermind-announces-market-eyeq-to-help-dealers-expand-their-customer-base-to-increase-sales

Hyland Software. https://en.wikipedia.org/wiki/Hyland_Software. Accessed 9 July 2020

IBM. In: Wikipedia. https://en.wikipedia.org/wiki/IBM. Accessed 22 July 2020.

IKEA. In: Wikipedia. https://en.wikipedia.org/wiki/IKEA. Accessed 22 July 2020.

In Cloudera Florida State University (2020): Boosting student success and powering decision-making through big data https://www.cloudera.com/about/customers/fsu.html Accessed 27 April 2020

Indiana University and Santa Barbara CC help students succeed with course-level. In: Tableau. https://www.tableau.com/solutions/customer/tale-two-universities-tableau-higher-education. Accessed 27 April 2020

Infection Control Today (2011), "X-Static Silver-Based Antimicrobial Fiber Technology Offers Protection", Retrieved from https://www.infectioncontroltoday.com/view/x-static-silver-based-antimicrobial-fiber-technology-offers-protection

Innovative Solutions & Development. In: LinkedIn. https://www.linkedin.com/company/innovative-solutions-development/. Accessed 9 July 2020

Inspiring classroom experiences. In: Smart Technologies. https://www.smarttech.com/en/education. Accessed 25 April 2020

International Finance Corporation, 2020a. The impact of COVID-19 on the Global Telecommunications Industry. Retrieved from https://www.ifc.org/wps/wcm/connect/industry_ext_content/ifc_external_corporate_site/infrastructure/resources/covid-19+impact+on+the+global+telecommunications+industry

International Labor Organization (2020), "COVID-19 and the Textiles, Clothing, Leather, and Footwear industries", Retrieved from https://www.ilo.org/sector/Resources/publications/WCMS_741344/lang%2D%2Den/index.htm

Investopedia, 2019. Financial Services Sector. Retrieved from https://www.investopedia.com/ask/answers/030315/what-financial-services-sector.asp

IoT Solution Development Services for Education. In: Integra Sources. https://www.integrasources.com/iot/iot-development-services-in-education/. Accessed 26 April 2020

Irawan J, Adriantantri E, Farid A (2017) RFID and IOT for Attendance Monitoring System. In: Matec Conferences. https://www.matec-conferences.org/articles/matecconf/pdf/2018/23/matecconf_icesti2018_01020.pdf. Accessed 26 April 2020

Is Pokémon Go Augmented Reality? In: Scientific American. https://www.scientificamerican.com/article/is-pokemon-go-really-augmented-reality/. Accessed 24 June 2020.

Jason B., 2019a Supervised and Unsupervised Machine Learning Algorithms. Retrieved from https://machinelearningmastery.com/supervised-and-unsupervised-machine-learning-algorithms/

Jeanologia (2020a) About Jeanologia

Jeanologia (2020b) EIM

Jeffry P., 2020. 10 Ways Banks and Credit Unions Are Using Virtual Reality. Retrieved from https://thefinancialbrand.com/68593/banks-credit-unions-finances-virtual-reality/

Jesus, A. (2018), "Artificial Intelligence in Industrial Automation – Current Applications", Retrieved from: https://emerj.com/ai-sector-overviews/artificial-intelligence-industrial-automation-current-applications/

Jiang F, Jiang Y, Zhi H, et al Artificial intelligence in healthcare: past, present and future Stroke and Vascular Neurology 2017a; DOI: https://doi.org/10.1136/svn-2017-000101

Jiang F, Zhi H, Jiang Y, Li H, Ma S, Wang Y, Dong Q, Shen H, Wang Y (2017b) Artificial Intelligence in Healthcare: Past, Present and Future Stroke, and Vascular Neurology. In: National Center for Biotechnology Information. https://www.ncbi.nlm.nih.gov/pmc/articles/PMC5829945/. Accessed 16 July 2020.

Jimmy. (2018). BMW Combines Car and Ride Sharing. Retrieved from SharedMobility: https://www.sharedmobility.news/bmw-combines-car-and-ride-sharing/

Jordan J (2019) Medtronic Launches the First Artificial Intelligence System for Colonoscopy at United European Gastroenterology Week 2019. In: Medtronic. http://newsroom.medtronic.com/news-releases/news-release-details/medtronic-launches-first-artificial-intelligence-system. Accessed 16 July 2020.

Kaspersky, "What is Biometrics", Retrieved from: https://usa.kaspersky.com/resource-center/definitions/biometrics

Kendris 2020. https://info.kendris.com/wealth-management/

Kenton W (2020) What is Green Tech? In: Investopedia. https://www.investopedia.com/terms/g/green_tech.asp. Accessed 31 July 2020.

Kolodny L. (2017) Meet Flippy, A Burger- Grilling Robot From Miso Robotics and CaliBurger. In: TechCrunch. https://techcrunch.com/2017/03/07/meet-flippy-a-burger-grilling-robot-from-miso-robotics-and-caliburger/. Accessed 22 July 2020.

Kovach N. Augmented Reality in Education. In: ThinkMobiles. https://thinkmobiles.com/blog/augmented-reality-education/. Accessed 9 July 2020.

Kramer S (2017) How Microsoft's HoloLens is Changing Medicine and Surgery. In: Futurum. https://futurumresearch.com/how-microsoft-hololens-is-changing-medicine-and-surgery/. Accessed 1 July 2020.

Kristen Korosec. (2018, July). BMW's Reach Now adds ride hailing to its car-sharing app. Retrieved from TechCrunch: https://techcrunch.com/2018/07/17/bmw-reachnow-ridesharing-car-sharing-app/

Kristen Korosec. (2020, March). Waymo suspends robotaxi service except for its truly driverless vehicles. Retrieved from TechiExpert: https://techcrunch.com/2020/03/17/waymo-suspends-robotaxi-service-except-for-its-truly-driverless-vehicles/

Kuebel, Hannes, and Ruediger Zarnekow. Evaluating Platform Business Models in the Telecommunications Industry via Framework-Based Case Studies of Cloud and Smart Home Service Platforms.

KYT, "BMW Announces Partchain Supply Blockchain for Tracking Automotive Parts", Retrieved from: https://siliconangle.com/2020/04/01/bmw-announces-partchain-supply-chain-blockchain-tracking-automotive-parts/

Lanier Worldwide, Inc. In Reference for Business. https://www.referenceforbusiness.com/history2/56/Lanier-Worldwide-Inc.html. Accessed 9 July 2020.

Levi Strauss & Co. In: Britannica. https://www.britannica.com/topic/Levi-Strauss-and-Co. Accessed 22 July 2020.

Levin A. (2019) Amazon Poised to Test Drone Deliveries Powered by Artificial Intelligence. In: TT Newsmaker. https://www.ttnews.com/articles/amazon-poised-test-drone-deliveries-powered-artificial-intelligence. Accessed 22 July 2020.

Liad C., 2020a. 4 Areas where AI Is Transforming the Telecom Industry in 2019. https://techsee.me/blog/artificial-intelligence-in-telecommunications-industry/

Linchpinseo, 2020. Challenges Facing The Real Estate Industry & Opportunities. Retrieved from https://linchpinseo.com/challenges-facing-the-real-estate-industry/ Retrieved from https://linchpinseo.com/challenges-facing-the-real-estate-industry/

LinkedIn Learning. In: Wikipedia. https://en.wikipedia.org/wiki/LinkedIn_Learning. Accessed 28 April 2020

LinkedIn, "A.T.E. Group", Retrieved from https://www.linkedin.com/company/a-t-e-enterprises-pvt-ltd-

LinkedIn, " Automotive Mastermind", Retrieved from: https://fr.linkedin.com/organization-guest/company/automotivemastermind-llc?challengeId=AQGJZQbnK4Am_gAAAXOQh4QZ-8bt3 QU7YcATMHbDWABRDJW0c76TN200CkfY5cSBELnJgPLy-dMl7Yqx4BmJYE0fOurTsT 9m1g&submissionId=c2385cb1-17a0-2516-4ea8-472eb0666b65

LinkedIn, " Tesla Motors", Retrieved from: https://www.linkedin.com/organization-guest/company/tesla-motors?challengeId=AQFn_BqAwh0EtAAAAXORizV3TbauMODoff9Ij31-A2yS5-eGMSLx5pOts0NAxB5pu_A8ewWkIMJ9Vn0JvcmJn5Q3b8A0ctvxAA&submission Id=805d7a4a-92af-2516-32d6-06d463ad7e02

LinkedIn, "Cooper Tire & Company", Retrieved from: https://www.linkedin.com/organization-guest/company/cooper-tire-&-rubber-company?challengeId=AQGwFUDB86smtQAAAXOR aSoCS9nZuc7xpb4sLco8q0z6nDLnPjf8SVeSzO_6OyINLwIH4CjkOvCjN6VH1pBAz4HbYi bWJ3XtZw&submissionId=caec4acf-8aad-2516-4f96-a4ef24baf523

LinkedIn, "General Motors", Retrieved from: https://eg.linkedin.com/company/general-motors

LinkedIn, "Nauto", Retrieved from: https://www.linkedin.com/company/nauto

LinkedIn, "Progress", Retrieved from: https://www.linkedin.com/organization-guest/company/progress-software?challengeId=AQHnlvDE7I1puAAAAXOU8UnnzSQB7ZHzLLWoUd1U 5rSyiyNuNmpsUnzf40HOsBXaj4RM5f7fYXf6DtC2Wi8kpNuu2y8xB5RIIw&submissionId =eba6efa3-6ee3-2516-6f25-1a91f0161512

Luan I. Data Mining Applications in Higher Education. In: SPSS https://www.spss.ch/upload/1122641492_Data%20mining%20applications%20in%20higher%20education.pdf. Accessed 28 April 2020

Lunas 3D, 2020. Interactive VR Tours Boosting Estate and Developers Market. Retrieved from https://www.lunas.pro/interactive-virtual-reality-tour.html

Lunden I (2020) Babylon Health is Building an Integrated, AI-Based Health App to Serve a City of 300k in England. In: Tech Crunch. https://techcrunch.com/2020/01/22/babylon-health-is-building-an-integrated-ai-based-health-app-to-serve-a-city-of-300k-in-england/. Accessed 1 July 2020.

Lupton (2013) The digitally Engaged Patient: Self-Monitoring and Self-Care in the Digital Health Era. In: Springlink. https://link.springer.com/article/10.1057/sth.2013.10. Accessed 16 July 2020.

Maccaron D (2020) 10 Relevant IoT Trends in 2020

Marelli, "Marelli Introduces Blockchain Technology to Bring Innovation into Automotive Supply Chain Management", Retrieved from: https://www.marelli.com/marelli-introduces-blockchain-technology-to-bring-innovation-into-automotive-supply-chain-management/

Margaret R., 2014a. What is biometrics? Retrieved from https://searchsecurity.techtarget.com/definition/biometrics

Marr B (2019a) The Amazing Ways Babylon Health is Using Artificial Intelligence to Make Healthcare Universally Accessible. In: Forbes. https://www.forbes.com/sites/bernard-marr/2019/08/16/the-amazing-ways-babylon-health-is-using-artificial-intelligence-to-make-healthcare-universally-accessible/#6d9010f23842. Accessed 1 July 2020.

Marr B (2019b) The Important Difference Between Virtual Reality, Augmented Reality and Mixed Reality. In: Forbes. https://www.forbes.com/sites/bernardmarr/2019/07/19/the-important-difference-between-virtual-reality-augmented-reality-and-mixed-reality/#797ac47435d3. Accessed 31 July 2020.

Marr, Bernard. "The Amazing Ways Telecom Companies Use Artificial Intelligence And Machine Learning." Forbes, Forbes Magazine, 3 Sept. 2019, www.forbes.com/sites/bernardmarr/2019/09/02/the-amazing-ways-telecom-companies-use-artificial-intelligence-and-machine-learning/#566a15b54cf6.

Marry B., 2019a. Introduction to Green Technology. Retrieved from https://www.thoughtco.com/introduction-to-green-technology-1991836

Martin P3 Integration. In: Green Hippo Support. https://support.green-hippo.com/article/24-martin-p3-integration. Accessed 24 June 2020.

Matchi.biz, 2020. Fintech innovations on Personal Financial Management (PFM). Retrieved from https://matchi.biz/news/808-fintech-innovations-on-personal-financial-management-pfm

Matshazi N (2019) PathAI Gets $75m For Their AI Pathology Algorithms. In: Healthcare Weekly. https://healthcareweekly.com/pathai-funding-round/. Accessed 16 July 2020.

McNeil S. (2018) Artificial Intelligence in the Classroom. In: Microsoft Education Blog. https://educationblog.microsoft.com/en-us/2018/03/artificial-intelligence-in-the-classroom/. Accessed 9 July 2020

Mearian L. (2019a), "What is Blockchain? The Complete Guide", Computer World. https://www.computerworld.com/article/3191077/what-is-blockchain-the-complete-guide.html

Mehdi N., 2017. Introduction to Machine Learning for Building Design and Construction. Retrieved from https://www.autodesk.com/autodesk-university/class/Introduction-Machine-Learning-Building-Design-and-Construction-2017

Meingast M, Roosta T, Sastry S (2006) Security and Privacy Issues with HealthcareInformation Technology. In: IEEE Xplore. https://ieeexplore.ieee.org/abstract/document/4463039. Accessed 16 July 2020.

Mercedes-Benz, "Vision AVTR – Inspired by Avatar", Retrieved from: https://www.mercedes-benz.com/en/vehicles/passenger-cars/mercedes-benz-concept-cars/vision-avtr/

Middleton, James. "Adventures in Nanotechnology." Telecoms.com, Https://Telecoms.com/, 10 Jan. 2015a, telecoms.com/22598/adventures-in-nanotechnology/.

Miliard M (2018) New Machine Learning Platform From Ciox Applies AI to Interoperability. In: Healthcare IT News. https://www.healthcareitnews.com/news/new-machine-learning-platform-ciox-applies-ai-interoperability. Accessed 16 July 2020.

Mire S. (2019) AI in Education Use Case #7: Gradescope. In: Disruptor Daily. https://www.disruptordaily.com/ai-in-education-use-case-7-gradescope/. Accessed 9 July 2020

Mitsubishi Electric, "Mitsubishi Electric Develops Robust Sensing for Autonomous Driving". Retrieved from: https://www.mitsubishielectric.com/sites/news/2019/pdf/0213-e.pdf

Moazed A. Platform Business Model. In: Applico. https://www.applicoinc.com/blog/what-is-a-platform-business-model/. Accessed 31 July 2020.

Munesh J., 2020a. Top 5 challenges & trends in the telecommunication industry in 2020. Retrieved from https://www.racknap.com/blog/top-5-challenges-trends-telecommunication-industry/

Nathan R., 2020a. Blockchain Explained. Retrieved from https://www.investopedia.com/terms/b/blockchain.asp

National Health Service. In: Wikipedia. https://en.wikipedia.org/wiki/National_Health_Service_(England). Accessed 16 July 2020.

Nauto, "Nauto Predictive Collision Alerts", Retrieved from: https://www.nauto.com/product/predictive-collision-alerts

Nawrat A (2018) 3D Printing in The Medical Field: Four Major Applications Revolutionizing The Industry. In: Verdict Medical Device. https://www.medicaldevice-network.com. Accessed 16 July 2020.

NHTSA, "Vehicle-to-Vehicle Communication", Retrieved from: https://www.nhtsa.gov/technology-innovation/vehicle-vehicle-communication

Nilsson N., Artificial Intelligence: A New Synthesis, ISBN-13: 978-1558604674, ISBN-10: 1558604677, 2009a https://books.google.com.eg/books?hl=en&lr=&id=Gt7gKuzL_8AC&oi=fnd&pg=PP2&dq=artificial+intelligence&ots=Q_dlSGWB7C&sig=3hnJ0jAlSBMXglAD9xQqJgQpOoQ&redir_esc=y#v=onepage&q=artificial%20intelligence&f=false

Nilsson N., Artificial Intelligence: A New Synthesis, ISBN-13: 978-1558604674, ISBN-10: 1558604677, 2009b https://books.google.com.eg/books?hl=en&lr=&id=Gt7gKuzL_8AC&oi

=fnd&pg=PP2&dq=artificial+intelligence&ots=Q_dlSGWB7C&sig=3hnJ0jAlSBMXglAD9x
QqJgQpOoQ&redir_esc=y#v=onepage&q=artificial%20intelligence&f=false

Noble Materials, "A solution for the Soft Surfaces Surrounding Healthcare Staff, Patients, and Families with Products Powered by X-Static Antimicrobial Technology", Retrieved from http://noblebiomaterials.com/xstatic-healthcare/

Notch. In: Green Hippo. https://www.green-hippo.com/about/partnerships/notch/. Accessed 24 June 2020.

NYU Langone Health, Helen L. and Martin S. Kimmel Pavilion. In: Ennead. http://www.ennead. com/work/nyulh-kimmel. Accessed 16 July 2020.

NYU's Tandon School of Engineering enrolls more women with the help of VR apps (2020). In: Google. https://edu.google.com/why-google/case-studies/nyu-tandon/?modal_active=none. Accessed 29 April 2020

OAL Group, "April Eye Label and Date Code Verification", Retrieved from https://connected. oalgroup.com/april-eye/

Our Brand. In: LG. https://www.lg.com/global/about-our-brand#overview Accessed 23 May 2020.

Our Company. In: Sam's Club. https://corporate.samsclub.com/our-company. Accessed 22 July 2020.

Outlast Technologies Inc., "Technology", Retrieved from http://www.outlast.com/en/technology/

Parkmobile, "Park. Phone. Go.", Retrieved from: http://www.parkmobile.co.uk

Partners. In: Amgen. https://www.amgen.com/partners/. Accessed 16 July 2020.

PathAI Case Study. In: Aptible. https://www.aptible.com/customers/pathai. Accessed 12 July 2020.

Pearson A. What is 3D Printing? In: Stratasys. https://www.stratasys.com/explore/article/what-is-3d-printing. Accessed 16 July 2020.

Penny J., 2018. How IoT Devices Enable Predictive Maintenance. https://www.buildings.com/ news/industry-news/articleid/21542/title/how-iot-devices-enable-predictive maintenance

Persado (2020), About Persado

Peter P., 2020. how much are fxpro spreads? are they good? https://brokerreview.net/ how-much-is-fxpro-spread-and-commission

Peter W., 2019. Augmented reality in building design and construction. Retrieved from https:// www.csemag.com/articles/augmented-reality-in-building-design-and-construction/

Plewa, Kat. "Top Projects Using 3D Printing for Telecommunication." Sculpteo, 3 Apr. 2019a, www.sculpteo.com/blog/2019/03/04/how-can-we-use-3d-printing-for-telecommunication/.

Pratap, A. (2018), "Obtaining a Competitive Advantage in the Automobile Industry", Retrieved from: https://notesmatic.com/2018/03/ obtaining-a-competitive-advantage-in-the-automobile-industry/

Pro Tools System. In: HowStuffWorks. https://entertainment.howstuffworks.com/pro-tools-software-hardware1.htm. Accessed 24 June 2020.

Progress, " Optimize Across the Entire Automotive Enterprise", Retrieved from: https://www. progress.com/datarpm/automotive

PTI (2019) Arvind Plans to Cut Carbon Emissions by 30 pc

Raaz N(2014) Current Challenges in the Global Textile Industry

Ranger S (2020a) What is IoT? In: ZDNET. https://www.zdnet.com/article/what-is-the-internet-of-things-everything-you-need-to-know-about-the-iot-right-now/. Accessed 31 July 2020.

Ranger S (2020b) What is IoT? In: ZDNET. https://www.zdnet.com/article/what-is-the-internet-of-things-everything-you-need-to-know-about-the-iot-right-now/. Accessed 31 July 2020.

Ranger S (2020c) What is the IoT? Everything You Need to Know About the Internet of Things Right Now. In: ZDNet. https://www.zdnet.com/article/what-is-the-internet-of-things-everything-you-need-to-know-about-the-iot-right-now/. Accessed 16 July 2020.

Rayven (2020a) About Rayvenhttps://www.rayven.io/about-us/ Accessed 19 May 2020

Rayven (2020b) Predictive Maintenance https://www.rayven.io/predictive-maintenance/ Accessed 19 May 2020

Redshift AutoDesk, 2019. About Redshift. Retrieved from https://www.autodesk.com/ redshift/about/

Rehfish M., 2020. Coronavirus: How it Impacts the Financial Services Sector. Retrieve from https://www.knowis.com/blog/coronavirus-how-it-impacts-the-financial-services-sector

Reiff N. (2019) What is SoftBank. In: Investopedia. https://www.investopedia.com/news/what-softbank/. Accessed 22 July 2020.

Ricci M (2017) Health Technology Firm Sensely has Partnered with the Mayo Clinic to Expand its Virtual Nurse Platform | Pharmaphorum. https://pharmaphorum.com/news/sensely-mayo-clinic-develop-virtual-doctor/. Accessed 1 July 2020.

Robotic Surgery. In: Cancer Treatment Centers of America. https://www.cancercenter.com/treatment-options/surgery/surgical-oncology/robotic-surgery. Accessed 1 July 2020.

Rosic A (2016a) What Is Blockchain Technology? A Step-By-Step Guide For Beginners. In: Block Geeks. https://blockgeeks.com/guides/what-is-blockchain-technology/. Accessed 31 July 2020.

Rosic A (2016b) What Is Blockchain Technology? A Step-By-Step Guide For Beginners. In: Block Geeks. https://blockgeeks.com/guides/what-is-blockchain-technology/. Accessed 31 July 2020.

Rouse M. Artificial Intelligence. In: Search Enterprise AI. https://searchenterpriseai.techtarget.com/definition/AI-Artificial-Intelligence. Accessed 31 July 2020.

Rouse M. Artificial Intelligence. In: Search Enterprise AI. https://searchenterpriseai.techtarget.com/definition/AI-Artificial-Intelligence. Accessed 31 July 2020.

Saavedra J. (2020) Educational Challenges and Opportunities of the Coronavirus (COVID-19) Pandemic. In: World Bank Blogs. https://blogs.worldbank.org/education/educational-challenges-and-opportunities-covid-19-pandemic. Accessed 9 July 2020

Sara Saunders. (2018, Aug). Betatype Case Study Illustrates Cost and Time Savings of Using 3D Printing to Fabricate Automotive Components. Retrieved from 3DPrint: https://3dprint.com/222506/betatype-automotive-case-study/

ScienceDirect, "Food and Beverage Industry", Retrieved from https://www.sciencedirect.com/topics/engineering/food-and-beverage-industry

ScienceDirect, "Food and Beverage Industry", Retrieved from https://www.sciencedirect.com/topics/engineering/food-and-beverage-industry

Scott Automation, "Automation and Robotics", Retrieved from https://www.scottautomation.com

Scott Automation, "Automation and Robotics", Retrieved from https://www.scottautomation.com

Scott Automation, "Robotic Beef Rib Cutting", Retrieved from https://www.scottautomation.com/assets/Uploads/Robotic-Beef-Rib-Cutting-Scribing-Scott2.pdf

Scott Automation, "Robotic Beef Rib Cutting", Retrieved from https://www.scottautomation.com/assets/Uploads/Robotic-Beef-Rib-Cutting-Scribing-Scott2.pdf

Sensely Overview. In: . https://www..com/organization/sense-ly#section-overview. Accessed 1 July 2020.

Sephora. In: Wikipedia. https://en.wikipedia.org/wiki/Sephora. Accessed 22 July 2020.

Sewbo, "Press Kit", Retrieved from http://www.sewbo.com/press/

Sharma, Ray. "Vodafone UK Launches AI-Based Chatbot Powered by IBM Watson." The Fast Mode, The Fast Mode, 17 Apr. 2017a, www.thefastmode.com/technology-solutions/10409-vodafone-uk-launches-ai-based-chatbot-powered-by-ibm-watson.

Sift. In: LinkedIn. https://www.linkedin.com/company/getsift/about/. Accessed 22 July 2020.

Silva (2020) 8K Resolution — Beyond 4K. In: Lifewire. https://www.lifewire.com/8k-resolution-beyond-4k- 1846844. Accessed 23 May 2020.

Singh, Vipin, et al. "Blockchain in Telecom: Which Companies Are Leading the Research?" GreyB, GreyB, 5 Feb. 2020a, www.greyb.com/companies-researching-blockchain-telecom-solutions/.

SmartLabel, "Welcome to SmartLabel", Retrieved from http://www.smartlabel.org

SmartLabel, "Welcome to SmartLabel", Retrieved from http://www.smartlabel.org

Smith A (2019) Medtronic Launches GI Genius. In: PharmaTimes. http://www.pharmatimes.com/news/medtronic_launches_gi_genius_1314109. Accessed 16 July 2020.

SODIC 2020. Human Developments. http://ir.sodic.com

SoftWear Automation, "A Fully Automated Sewing Work line Built to Scale Sewn Goods Manufacturing", Retrieved from http://softwearautomation.com/products/

SPSS Inc. In Wikipedia. https://en.wikipedia.org/wiki/SPSS_Inc. Accessed 28 April 2020

Srikanth. (2017, Aug). How BMW Uses Artificial Intelligence And Big Data To Design And Build Cars Of Tomorrow. Retrieved from TechiExpert: https://www.techiexpert.com/bmw-uses-artificial-intelligence-big-data-design-build-cars-tomorrow/

Stanley G., 2020a. What Is a Platform?. Retrieved from https://www.lifewire.com/what-is-a-platform-4155653

Stephen G., 2019. AI in banking examples you should know. Retrieved from KNOW https://builtin.com/artificial-intelligence/ai-in-banking

Steve R., 2019a What is the IoT? Everything you need to know about the Internet of Things right now. Retrieved from https://www.zdnet.com/article/what-is-the-internet-of-things-everything-you-need-to-know-about-the-iot-right-now/

Steve R., 2019b What is the IoT? Everything you need to know about the Internet of Things right now. Retrieved from https://www.zdnet.com/article/what-is-the-internet-of-things-everything-you-need-to-know-about-the-iot-right-now/

Steve R., 2019c What is the IoT? Everything you need to know about the Internet of Things right now. Retrieved from https://www.zdnet.com/article/what-is-the-internet-of-things-everything-you-need-to-know-about-the-iot-right-now/

Steve R., 2019d What is the IoT? Everything you need to know about the Internet of Things right now. Retrieved from https://www.zdnet.com/article/what-is-the-internet-of-things-everything-you-need-to-know-about-the-iot-right-now/

Systems innovation 2020. Platform Technologies. Retrieved from https://systemsinnovation.io/platforms-technologies/

Technopedia (2020) Definition of Platform

Telit, "Improve Food Quality, Operational Efficiency, and Compliance", Retrieved from https://www.telit.com/industries-solutions/retail/food-and-beverage/

Telit, "Improve Food Quality, Operational Efficiency, and Compliance", Retrieved from https://www.telit.com/industries-solutions/retail/food-and-beverage/

Tender Tiger 2020. https://gammon.tendertiger.com/quicksearch.aspx?st=qs&SerCat=7&SerText=Building%20Construction%20Department&tt=&si=&tenders=Building%20Construction%20Department+tenders

Tesla, "Gigafactory", Retrieved from: https://www.tesla.com/gigafactory

The Changing Oncology Landscape. In: IBM. https://www.ibm.com/watson-health/oncology-and-genomics?mhsrc=ibmsearch_a&mhq=Watson%20for%20Cancer%20care. Accessed 16 July 2020.

The economical parliament, 2019. http://parliamenteconomic.com/ما-هي-خدمة-pay-التي-تقدمها-المصرية-الاتصال/

The Trust Chain Initiative. In: Trust Chain Jewelry. https://www.trustchainjewelry.com. Accessed 22 July 2020.

Thinkster Math. In: LinkedIn. https://www.linkedin.com/company/prazas-coaxis-services-inc/about/. Accessed 9 July 2020

Thomas M (2020) 15 Examples of Machine Learning in Healthcare That Are Revolutionizing Medicine. In: Builtin. https://builtin.com/artificial-intelligence/machine-learning-healthcare. Accessed 16 July 2020.

Toray Industries, "Together, We are the New Green", Retrieved from https://www.andplus.toray/en/

Trend Forecasting. In: Heuritech. https://www.heuritech.com/trend-forecasting/. Accessed 22 July 2020.

Understanding the sector impact of COVID-19 Media & Entertainment. In: Deloitte. https://www2.deloitte.com/global/en/pages/about-deloitte/articles/covid-19/covid-19-impact-on-media-and-entertainment-companies.html. Accessed 18 May 2020.

University of Zagreb University Computing Centre SRCE. https://moodle.com/wp-content/uploads/2019/07/University-of-Zagreb_Casestudy.pdf. In: Moodle. Accessed 29 April 2020

Vaidya A (2018) Northwestern Memorial Healthcare, Ciox Join Forces. In: Health IT. https://www.beckershospitalreview.com/healthcare-information-technology/northwestern-memorial-healthcare-ciox-join-forces.html. Accessed 16 July 2020.

Veo Robot, "Veo FreeMove: Bringing Together Humans and Industrial Robots for Flexible Manufacturing", Retrieved from: https://www.veobot.com/blog/2019/11/11/veo-freemove-bringing-together-humans-and-industrial-robots-for-flexible-manufacturing

Vincent, J. (2016), " A RoboGlove Designed by NASA and GM will Help Factory Workers Get a Grip", Retrieved from: https://www.theverge.com/circuitbreaker/2016/7/6/12105074/nasa-gm-power-glove-tech

Vodafone UK Launches IoT Heat Sensor to Help Coronavirus Fight. Geospatial World, 5 May 2020, www.geospatialworld.net/news/vodafone-uk-launches-iot-heat-sensor-to-help-coronavirus-fight/.

Wake Forest Institute for Regenerative Medicine. In: Wikipedia. https://en.wikipedia.org/wiki/Wake_Forest_Institute_for_Regenerative_Medicine. Accessed 16 July 2020.

Walk-Morris T. (2020) ASOS Debuts AR Tool For Online Shoppers. In: Retail Dive. https://www.retaildive.com/news/asos-debuts-ar-tool-for-online-shoppers/577679/. Accessed 22 July 2020.

Walmart. In: Wikipedia. https://en.wikipedia.org/wiki/Walmart. Accessed 22 July 2020.

Warren, T. (2017), "Ford is Using Microsoft's HoloLens to Design Cars in Augmented Reality", Retrieved from: https://www.theverge.com/2017/9/21/16343354/microsoft-hololens-ford-augmented-reality

Wasserman T. 10 Technologies That Can Change Retail Forever. In: CMO. https://cmo.adobe.com/articles/2017/10/10-technologies-helping-to-overhaul-the-retail-experience.html#gs.be7xlq. Accessed 22 July 2020.

Waymo, "Technology", Retrieved from: https://waymo.com/tech/

WeAccept 2020. https://www.weaccept.co

Wearable X, "Posture Monitoring and Vibrational Guidance", Retrieved from https://www.wearablex.com/pages/how-it-works

What is 3D Printing? In: Stratasys. https://www.stratasys.com/explore/article/what-is-3d-printing. Accessed 31 July 2020.

What is 3D Printing? In: Stratasys. https://www.stratasys.com/explore/article/what-is-3d-printing. Accessed 31 July 2020.

What Is Biometric Security: How It Works. In: Kaspersky. https://www.kaspersky.co.za/resource-center/definitions/biometrics. Accessed 31 July 2020.

What Is Biometric Security: How It Works. In: Kaspersky. https://www.kaspersky.co.za/resource-center/definitions/biometrics. Accessed 31 July 2020.

What Is Digitalization. In: IGI Global. https://www.igi-global.com/dictionary/it-strategy-follows-digitalization/7748. Accessed 31 July 2020.

What Is Digitalization. In: IGI Global. https://www.igi-global.com/dictionary/it-strategy-follows-digitalization/7748. Accessed 31 July 2020.

What is Nanotechnology? In: National Nanotechnology Initiative. https://www.nano.gov/nanotech-101/what/definition. Accessed 16 July 2020.

What is Nanotechnology? In: National Nanotechnology Initiative. https://www.nano.gov/nanotech-101/what/definition. Accessed 31 July 2020.

What is Nanotechnology? In: National Nanotechnology Initiative. https://www.nano.gov/nanotech-101/what/definition. Accessed 31 July 2020.

What We're About: About Domino's. In: Domino's. https://biz.dominos.com/web/public/about. Accessed 22 July 2020.

What We're About: About Domino's. In: Domino's. https://biz.dominos.com/web/public/about. Accessed 22 July 2020.

When museums became virtual. In: Inexhibit. https://www.inexhibit.com/case-studies/virtual-museums- part-1-the-origins/ Accessed 18 May 2020.

Williams, M., "What's an Automated Kitchen", Retrieved from https://www.herox.com/blog/352-robot-chefs-and-automated-kitchens

Williams, M., "What's an Automated Kitchen", Retrieved from https://www.herox.com/blog/352-robot-chefs-and-automated-kitchens

Wingard L., 2020. Top 10 Banking Industry Challenges — And How You Can Overcome Them. Retrieved from https://global.hitachi-solutions.com/blog/top-10-challenges-banking-financial-organizations-can-overcome

Wong W (2018) Innovative Hospitals Tap Automation to Streamline Patient Care. In: Health Tech. https://healthtechmagazine.net/article/2018/11/innovative-hospitals-tap-automation-streamline-patient-care. Accessed 16 July 2020.

XYHT, 2020. Hidden Infrastructure in 3D: Visualizing with AR. Retrieved from https://www.xyht.com/constructionbim/visualizing-hidden-infrastructure-in-3d/

Company Snapshot. In: Business of Fashion. https://www.businessoffashion.com/organisations/zara. Accessed 22 July 2020.

(2017) Introducing Hands-Free Check-In For Your Favorite Smart Device. In: United. https://hub.united.com/united-recipes-onboard-snack-2645937941.html. Accessed 30 June 2020.

(2020) Company Overview. In: United. https://ir.united.com/company-information/company-overview. Accessed 30 June 2020.

6 Companies Using Blockchain to Change Travel. In: Investopedia. https://www.investopedia.com/news/6-companies-using-blockchain-change-travel-0/. Accessed 30 June 2020.

About Avvio. In: Avvio. https://www.avvio.com/about-avvio/. Accessed 30 June 2020

AI in the Travel and Tourism Industry - Current Applications. In: Emerj. https://emerj.com/ai-sector-overviews/ai-travel-tourism-industry-current-applications/. Accessed 30 June 2020

Airbnb (2020), "2020 Airbnb Update". Retrieved from: https://news.airbnb.com/2020-update/

Allianz Business Insights (2017), "FlightBeat: Smart Airplane Seats Capable of Detecting Stress". https://allianzpartners-bi.com/news/flightbeat-smart-airplane-seats-capable-of-detecting-stress-b505-333d4.html

Atlantis, the Palm - VR Panovideo Application. In: Artworks Smartphone Application. http://www.arworks.com/en/portfolio-item/atlantis-the-palm-vr-panovideo-application/. Accessed 30 June 2020

Atlantis, The Palm Dubai. In: Circle One Studios. https://www.circleonestudios.com/360/Atlantis/. Accessed 30 June 2020.

Avvio's AI Booking Engine Powers New Chatbot Skill for Hotel Direct. In: Phocus Wire. https://www.phocuswire.com/Avvios-AI-booking-engine-powers-new-chatbot-skill-for-hotel-direct. Accessed 30 June 2020.

British Airways (2019), "British Airways Factsheet", Media Centre. https://mediacentre.britishairways.com/factsheets/details/86/Factsheets-3/33

Burgess M. (2018), "What is the Internet of Things? WIRED Explains", WIRED. https://www.wired.co.uk/article/internet-of-things-what-is-explained-iot

Business Background of Hilton Hotels. In: UK Essays. https://www.ukessays.com/essays/marketing/business-background-of-hilton-hotels-corporation-marketing-essay.php. Accessed 30 June 2020.

Chloe. (2019, March). What Are The Top Challenges Facing the Travel Industry In 2019? Retrieved from SiteVisibility: https://www.sitevisibility.co.uk/blog/2019/03/08/top-challenges-facing-travel-industry-2019/#gref. Accessed 9 March 2021

Computer Business Review (2020), "What is IBM", Retrieved from: https://www.cbronline.com/what-is/what-is-ibm-4950406/. Accessed 9 March 2021

COVID-19: PUTTING PEOPLE FIRST. Retrieved from UNWTO: https://www.unwto.org/event-suspended. Accessed 9 March 2021

Delta News Hub (2020), "Corporate Stats and Facts", Retrieved from: https://news.delta.com/corporate-stats-and-facts. Accessed 9 March 2021

Expedia Group, Inc. Modern Slavery Statement. In: Expedia Group. https://www.expediagroup.com/about/msa-statement/. Accessed 30 June 2020.

Grigonis H. K. (2017), "Like a Siri for Travel, HelloGbye is a New Streamlined Travel-planning App", Digital Trends. https://www.digitaltrends.com/mobile/hellogbye-launches-on-ios/

HelloGbye, "Improve Your Traveler Experience". Retrieved from: http://www.hellogbye.com. Accessed 9 March 2021

Hilton and IBM Pilot "Connie," The World's First Watson-Enabled Hotel Concierge. In: Hilton. https://newsroom.hilton.com/corporate/news/hilton-and-ibm-pilot-connie-the-worlds-first-watsonenabled-hotel-concierge. Accessed 30 June 2020.

Hopper, "Never Overpay for Travel Again". Retrieved from: https://www.hopper.com

Houser K. (2020), "Airbnb Claims its AI can Predict Whether Guests are Psychopaths". https://futurism.com/the-byte/airbnb-ai-predict-psychopaths

How Blockchain Technology is Transforming the Travel Industry. In: Refine. https://www.revfine.com/blockchain-technology-travel-industry/. Accessed 30 June 2020.

Kamps H. J. (2016), "Polarsteps Automates your Travel Blogging". https://techcrunch.com/2016/03/22/polarsteps-travel-blog/

Kang, Bomi; Brewer, Kathleen Pearl; and Bai, Billy (2007) "Biometrics for Hospitality and Tourism: A New Wave of Information Technology," Hospitality Review: Vol. 25: Iss. 1, Article 1. Available at: https://digitalcommons.fiu.edu/hospitalityreview/vol25/iss1/1

KLM, "KLM Company Profile", Retrieved from: https://www.klm.com/travel/nl_en/corporate/company_profile.htm

Koines TJ (2018), "Advancing the Design of NavCog", IBM. https://www.ibm.com/blogs/age-and-ability/2018/09/11/advancing-the-design-of-navcog/

LinkedIn, "Peek: About us". Retrieved from: https://www.linkedin.com/company/peek-com

Marr B., "How Airbnb Uses Big Data in Practice". Retrieved from: https://www.bernardmarr.com/default.asp?contentID=708

Marriott International (2018), " Joint Venture of Alibaba Group and Marriott International Trials Facial Recognition Check-In Technology", New Center. https://news.marriott.com/news/2018/07/11/joint-venture-of-alibaba-group-and-marriott-international-trials-facial-recognition-check-in-technology

Mathieu (2019), "5 Ways Artificial Intelligence (AI)is Changing Travel and Tourism". Retrieved from: https://www.bocasay.com/5-ways-artificial-intelligence-is-changing-travel-and-tourism/

Mearian L. (2019b), "What is Blockchain? The Complete Guide", Computer World. https://www.computerworld.com/article/3191077/what-is-blockchain-the-complete-guide.html

Peek, "Sell More Experiences. Spend Less Time Doing it". Retrieved from: https://www.peek.com/pro/

Perez S. (2019), "AI-Based Travel App Hopper Expands Price Monitoring to Hotels Global". https://techcrunch.com/2019/06/20/a-i-based-travel-app-hopper-expands-price-monitoring-to-hotels-Global/

Polarsteps, "Make Your Way in the World". Retrieved from: https://www.polarsteps.com

Refine, "How Blockchain Technology is Transforming the Travel Industry". Retrieved from: https://www.revfine.com/blockchain-technology-travel-industry/

Refine, "How the Internet of Things (IoT) can Benefit the Travel Industry". Retrieved from: https://www.revfine.com/internet-of-things-travel-industry/

Rouse M. (2010), "Artificial Intelligence", Search Enterprise AI. https://searchenterpriseai.techtarget.com/definition/AI-Artificial-Intelligence

Rouse, M. (2012, July). What is Big Data Analytics and Why is it important? Retrieved from SearchBusinessAnalytics: https://searchbusinessanalytics.techtarget.com/definition/big-data-analytics

Softbank Robotics, "About Softbank Robotics", Retrieved from: https://www.softbankrobotics.com/emea/en/company

Szondy D. (2014), "British Airways Tests "Happiness Blanket", New Atlas. https://newatlas.com/british-airways-tests-happiness-blanket/32799/

Tourism Industry: Everything you need to know about Tourism. Retrieved from REVFINE: https://www.revfine.com/tourism-industry/

Travel Massive, "Utrip: Travel Planning", Retrieved from: https://travelmassive.com/utrip

Travelmate Robotics, "Travelmate: a Fully Autonomous Suitcase and Robot", Retrieved from: https://travelmaterobotics.com

What is the tourism industry? Retrieved from MarketWidth: http://www.market-width.com/Tourism-Industry.htm

Wind Horse Tour, "Beijing Capital International Airport". Retrieved from: https://windhorsetour.com/beijing-attraction/beijing-capital-international-airport

3D Printing Egypt - On-Demand 3D Printing in Egypt. ETBA3LY, www.etba3ly3d.com/.

About – Ubitquity. Ubitquity, 2021, www.ubitquity.io/about.

AIST: About AIST. AIST, 2021, www.aist.go.jp/aist_e/about_aist/index.html. Accessed 8th of Feb 2021

Brass, Jacob. 7 Impressive Construction Tech Innovations for 2020. Gear flow Blog, 17 Aug. 2020, gearflow.com/blog/2020-construction-tech-watchlist.

Clearpath Robotics. Husky UGV - Outdoor Field Research Robot by Clearpath. Clearpath Robotics, 21 Dec. 2020, clearpathrobotics.com/husky-unmanned-ground-vehicle-robot.

Dozr. Construction Equipment Rental - Search and Book Online I DOZR. Dozr, 2021, dozr.com. Accessed 7th of Feb 2021

Engineering, Construction & Project Management. Bechtel Corporate, 2021, www.bechtel.com.

Fraunhofer in Europe. Fraunhofer, 2020, www.fraunhofer.de/en/institutes/international/europe/italy.html.

Hearns, Alicia. Newly Re-Engineered SmartRock™ Concrete Sensor Launches. Giatec Scientific Inc., 15 Sept. 2020, www.giatecscientific.com/company-news/giatecs-newly-re-engineered-smartrock-concrete-sensor-launches-with-dual-temperature-functionality/.

Hill, Trevor. Ubitquity Used to Test Pilot Blockchain Land Registry in Brazil. Bitsonline, 27 Jan. 2018, bitsonline.com/ubitquity-test-blockchain-land.

Improving Safety on Construction Sites with XR. Bechtel Corporate, 2021, www.bechtel.com/blog/innovation/november-2018/using-extended-reality-improve-safety-construction/#:%7E:text=For%20example%2C%20Bechtel%20is%20currently,interest%20throughout%20the%20construction%20site.

Klapty. Klapty, www.klapty.com. Accessed 31 Jan. 2021.

"KTH Royal Institute of Technology." Top Universities, 16 July 2015, www.topuniversities.com/universities/kth-royal-institute-technology.

Lidija Grozdanic for Archipreneur.com. The Top 5 Virtual Reality and Augmented Reality (AR)Apps for Architects. ArchDaily, 17 Sept. 2017, www.archdaily.com/878408/the-top-5-virtual-reality-and-augmented-reality-apps-for-architects.

Maturix. Sensohive, 6 Nov. 2020, sensohive.com/maturix. accessed: 15th of Feb 2021

MTWO Go Live in 48Hrs. MTWO, 2021, www.mtwocloud.com/mtwo-go-live-in-48hrs. Accessed 15th of Feb 2021

Obudho, Brian. 5 Biggest Companies Building 3D Printed Houses. All3DP, 9 Oct. 2019, all3dp.com/2/2019-best-companies-building-3d-printed-houses.

Orbit K. Sensohive, 26 Jan. 2021, sensohive.com/sensors/orbitk.

Oyj, Kone. KONE Revolutionizes Elevator Maintenance with New Customizable KONE Care(TM) Service Offering and 24/7 Connected Services. GlobeNewswire News Room, 8 Feb. 2017, www.globenewswire.com/news-release/2017/02/08/914900/0/en/KONE-revolutionizes-elevator-maintenance-with-new-customizable-KONE-Care-TM-service-offering-and-24-7-Connected-Services.html.

Pemberton, Kelly. 8 Sustainable Innovations That Are Shaping the Construction Sector. Green and Prosperous, 5 Mar. 2019, www.greenandprosperous.com/blog/8-sustainable-innovations-shaping-the-construction-sector.

Phillips, Zachary. Robot Roundup: 5 Recent Innovations in Construction Tech. Construction Dive, 1 July 2020, www.constructiondive.com/news/robot-roundup-5-recent-innovations-in-construction-tech/580902.

Stambol. Augmented Reality (AR)Tours for Real Estate & Beyond. Stambol, 7 July 2020, www.stambol.com/2020/07/06/augmented-reality-tours-for-real-estate-beyond.

Telit, et al. IoT Applications in Construction. IoT For All, 14 July 2020, www.iotforall.com/iot-applications-construction.

Ubitquity.io. 2021. [online] Available at: https://www.ubitquity.io/wp-content/uploads/2020/01/UBITQUITY-CaseStudy.pdf [Accessed 14 February 2021]

About the Author: "The State of Biometrics Solutions: Use Cases and Advances." Aware, 20 Jan. 2020, www.aware.com/blog-state-of-biometrics-solutions/.

Ahmed, Faran, et al. "ICT and Renewable Energy: a Way Forward to the Next Generation Telecom Base Stations." Telecommunication Systems, Springer US, 1 Jan. 1970, link.springer.com/article/10.1007/s11235-016-0156-4.

Banerjee, Ari. "Big Data & Advanced Analytics in Telecom: A Multi-Billion Revenue Opportunity." Huawei, 2016b, www.huawei.com/ilink/en/download/HW_323807.

Bardi J., 2019c What is Virtual Reality? [Definition and Examples]. Retrieved from https://www.marxentlabs.com/what-is-virtual-reality/

Burch, Aaron. "The Top 10 Companies Working on Education in Virtual Reality and Augmented Reality." Touchstone Research, 29 Sept. 2016b, touchstoneresearch.com/the-top-10-companies-working-on-education-in-virtual-reality-and-augmented-reality/.

Car Connectivity Services by Borgward & Orange. Ericsson.com, 9 Mar. 2020b, www.ericsson.com/en/cases/2019/orange-and-borgward.

Carol M., 2017b. Big Data Opportunities for Telecommunications. Retrieved from https://mapr.com/blog/big-data-opportunities-telecommunications/

Connor C., 2020b. How are Telecoms Using the Internet of Things (IoT)? https://www.sdxcentral.com/5g/iot/definitions/telecom-using-iot/

International Finance Corporation, 2020b. The impact of COVID-19 on the Global Telecommunications Industry. Retrieved from https://www.ifc.org/wps/wcm/connect/industry_ext_content/ifc_external_corporate_site/infrastructure/resources/covid-19+impact+on+the+global+telecommunications+industry

Jason B., 2019b Supervised and Unsupervised Machine Learning Algorithms. Retrieved from https://machinelearningmastery.com/supervised-and-unsupervised-machine-learning- algorithms/

Jiang F, Jiang Y, Zhi H, et al Artificial intelligence in healthcare: past, present, and future Stroke and Vascular Neurology 2017c; DOI: https://doi.org/10.1136/svn-2017-000101

Kuebel, Hannes, and Ruediger Zarnekow. Evaluating Platform Business Models in the Telecommunications Industry via Framework-Based Case Studies of Cloud and Smart Home Service Platforms.

Kuebel, Hannes, and Ruediger Zarnekow. Evaluating Platform Business Models in the Telecommunications Industry via Framework-Based Case Studies of Cloud and Smart Home Service Platforms.

Liad C., 2020b. 4 Areas where AI Is Transforming the Telecom Industry in 2019. https://techsee.me/blog/artificial-intelligence-in-telecommunications-industry/

Margaret R., 2014b. What is biometrics? Retrieved from https://searchsecurity.techtarget.com/definition/biometrics

Marr, Bernard. "The Amazing Ways Telecom Companies Use Artificial Intelligence And Machine Learning." Forbes, Forbes Magazine, 3 Sept. 2019d, www.forbes.com/sites/bernardmarr/2019/09/02/the-amazing-ways-telecom-companies-use-artificial-intelligence-and-machine-learning/#566a15b54cf6.

Marry B., 2019b. Introduction to Green Technology. Retrieved from https://www.thoughtco.com/introduction-to-green-technology-1991836

Middleton, James. "Adventures in Nanotechnology." Telecoms.com, Https://Telecoms.com/, 10 Jan. 2015b, telecoms.com/22598/adventures-in-nanotechnology/.

Munesh J., 2020b. Top 5 challenges & trends in the telecommunication industry in 2020. Retrieved from https://www.racknap.com/blog/top-5-challenges-trends-telecommunication-industry/

Nathan R., 2020b. Blockchain Explained. Retrieved from https://www.investopedia.com/terms/b/blockchain.asp

Plewa, Kat. "Top Projects Using 3D Printing for Telecommunication." Sculpteo, 3 Apr. 2019b, www.sculpteo.com/blog/2019/03/04/how-can-we-use-3d-printing-for-telecommunication/.

Sharma, Ray. "Vodafone UK Launches AI-Based Chatbot Powered by IBM Watson." The Fast Mode, The Fast Mode, 17 Apr. 2017b, www.thefastmode.com/technology-solutions/10409-vodafone-uk-launches-ai-based-chatbot-powered-by-ibm-watson.

Singh, Vipin, et al. "Blockchain in Telecom: Which Companies Are Leading the Research?" GreyB, GreyB, 5 Feb. 2020b, www.greyb.com/companies-researching-blockchain-telecom-solutions/.

Stanley G., 2020b. What Is a Platform?. Retrieved from https://www.lifewire.com/what-is-a- platform-4155653

Steve R., 2019e What is the IoT? Everything you need to know about the Internet of Things right now. Retrieved from https://www.zdnet.com/article/what-is-the-internet-of-things-everything- you-need-to-know-about-the-iot-right-now/

Vodafone UK Launches IoT Heat Sensor to Help Coronavirus Fight. Geospatial World, 5 May 2020, www.geospatialworld.net/news/vodafone-uk-launches-iot-heat-sensor-to-help- coronavirus-fight/

Index

Printed in the United States
by Baker & Taylor Publisher Services